FUNDAMENTALS OF
SOIL SCIENCE

Related Books

• Agriculture Chemical Users Manual	*Geoff Cowles*
• Agrotechnology for Dryland Farmaing *2nd Revised Ed*	*A.M. Dhopte*
• Allelopathy in Soil Sickness	*S.S. Narwal*
• Biofertilizer Technology	*S. Kannaiyan*
• Comprehensive assessment of soil health *3rd Ed*	*SIPS*
• Crop Production in Salt Affected Soils	*I.C. Gupta*
• Crop Production in Waterlogged Saline Soils	*S.K. Gupta*
• Crop Protection Strategies under Climate Change Scenarios	*P.P. Reddy*
• Drainage Engineering: Principles and Practices	*S.K. Gupta*
• Dryland Resources and Technology (Vol. 8)	*L.L. Somani*
• Dryland Technology *2nd Ed*	*M.L. Jat*
• Fertilizer Manual	*IFDC*
• Fundamentals of Soil Science	*A. Rathinasamy*
• Genesis and Management of Sodic (alkali) Soils	*S.K. Gupta*
• Handbook of Saline and Alkali Soils Diagnosis Reclamation and Management	*S.K. Gupta*
• Indiras Objective Agriculture	*R.L. Arya*
• Manual on Fertilizer Distribution**	*FAO*
• Methods of Analysis for Soils of Arid and Semi Arid Regions**	*I.I. Bashour*
• Methods of Soil Analysis: Microbiological and Biochemical Properties Part 2**	*R.W. Weaver*
• Nanofertilizers: Challenges and Prospects**	*J.C. Tarafdar*
• Nanotechnology	*J.C. Tarafdar*
• New Vistas of Organic Farming *2nd Ed*	*Mukund Joshi*
• Organic Farming Theory & Practice	*S.P. Palaniappan*
• Rainwater Management Theory and Practice	*M.L. Jat*
• Salinity Tolerance in Plants: Methods, Mechanisms and Management *2nd Ed*	*B.K. Garg*
• Sasya Vigyan Ke Sidhant (Principles of Agronomy) (Hindi)	*R.L. Arya*
• Soil and Plant Analysis	*C.S. Piper*
• Soil and Water Conservation in Semi-Arid Areas	*N.W. Hudson*
• Soil Chemical Analysis	*M.L. Jackson*
• Sustainable Agriculture	*R.S. Meena*
• Sustainable Dryland Farming**	*A.M. Dhopte*
• The Use of Saline Waters for Crop Production	*FAO*
• Tillage and Crop Production *2nd Ed*	*N.R. Das*
• Water Measurement Manual *3rd Ed*	*USDI*

FUNDAMENTALS OF
SOIL SCIENCE

Dr. R. L. Arya
M.Sc. & Ph.D., Agronomy,
I.A.R.I., PUSA, New Delhi
Ex. Senior Scientist, ICAR-CTRI, Rajahmundry (AP)

Dr. Khalil Khan
M.Sc. (Ag.) Ph.D., Soil Science,
Assistant Professor/Scientist
Directorate o Research
C.S. Azad University of Ag.
Tech Kanpur (UP)

SCIENTIFIC
PUBLISHERS

Published by
SCIENTIFIC PUBLISHERS (INDIA)
5 A, New Pali Road, P.O. Box 91
Jodhpur 342 001 (INDIA)
E-mail: info@scientificpub.com
Website: http://www.scientificpub.com

Print 2025

ISBN: 978-93-87741-2-63
eISBN: 978-93-87991-81-1

Printed in India

Contents

1. Introduction 1
2. Soils 4
3. Soil formation 9
4. Nutrient cycle 14
5. Soil organic matter 15
6. Soil water 21
7. Physical properties of soil 25
8. Chemical properties of soil 29
9. Biological properties of soil 31
10. Ion exchange 32
11. Electrical conductivity of soil 35
12. Redox potential 36
13. Plant nutrition 39
14. Soil fertility and productivity 45
15. Transformation of nutrients 48
16. Assimilation of nutrients 53
17. Mixed fertilizers 55
18. Organic fertilizers 58
19. Green manuring 60
20. Method and time of fertilizer application 63
21. Enhancing fertilizer use efficiency 65
22. Classification of fertilizers 66
23. Functions of essential plant nutrients 67
24. Deficiency symptoms of plant nutrient 71
25. Fate of plant nutrient 72
26. Soil erosion 76
27. Soil reaction 80
28. Soils of India 85
29. Classification of soil groups 88
30. Saline and alkali soils 89

31. Acid soils 93
32. Land capability classification 95
33. Soil nutrient mineralisation 97
34. Integrated plant nutrient management 100
35. Organic farming 102
36. Biofertilizers 105
37. Soil and water pollution 114
38. Quality of fertilizer 119
39. Indian fertilizer policy 122
40. The fertilizer (control) order, 1985 127
41. Important definitions of soil science 145

SOIL SCIENCE

1. INTRODUCTION

Soil science is the study of soil as a natural resource on the surface of the earth including soil formation, and mapping; physical, chemical, biological, and fertility properties of soils; and these properties in relation to the use and management of soils.

A study of soil science includes:

(i) **Soil geology:** The study of the geological materials from which the soil is derived and its process of formation.

(ii) **Soil chemistry:** The study of the chemical components of the soil, their interaction with one another and the effects of the chemical environment of the soil.

(iii) **Soil physics:** The study of the effects of physical laws on the evolution of chemical properties of soils.

(iv) **Soil biology:** The study of the effect of plants, animal and soil microorganisms on the evolution, chemical composition and physical condition of the soil.

There are two basic concepts of soil:

Sometimes terms which refer to branches of soil science, such as pedology edaphology

soil science.

1. **Pedology** is a Greek word where, Pedon means Soil/Earth and Logos means discourse/study.

According to pedology, soil is a natural body or entity and is a biochemically weathered and synthesized product of nature.

- Pedologist studies the origin of soil, classifies the soil and describes the soil in its natural environment.
- Pedology is the study of soil in its natural setting.

2. **Edaphology** is a Greek word where Edaphos means soil/ground and logos means discourse.

- To study the soil from the stand point of higher plants is known as edaphology.
- According to edaphology soil is the natural habitat for plants.
- Edaphology is the study of soil in relation to soil-dependent uses.
  - Both branches apply a combination of soil physics, soil chemistry and soil biology.
  - Due to the numerous interactions between the biosphere, atmosphere and hydrosphere that are hosted within the pedosphere, more integrated and less soil-centric concepts are also valuable.
  - Many concepts essential to understanding of soil come from individuals not identifiable strictly as soil scientists.
  - This highlights the interdisciplinary nature of soil concepts.
  - Dokuchaev, a Russian geologist and Soil Scientist, considers the soil as a natural body having its own genesis and its own history of development, a body with complex and multiform processes taking place within it. The soil is considered as different from bedrock. The latter becomes soil under the influence of a series of soil-formation factors (climate, vegetation, parent material, relief and age). According to him, soil should be called the "daily" or outward horizons of rocks regardless of the type; they are changed naturally by the common effect of water, air and various kinds of living and dead organisms.
  - Soil may be defined as: "the different forms of earth on the surface of the rocks, formed by the breaking down or weathering of rocks".
  - Dokuchaev's late 19th century soil concept developed in the 20th century to one of soil as earthy material that has been altered by living processes. A corollary concept is that soil without a living component is simply a part of earth's outer layer.
  - Further refinement of the soil concept is occurring in view of an appreciation of energy transport and transformation within soil. The term is popularly applied to the material on the surface of the Earth's moon and Mars, a usage acceptable within a portion of the scientific community.
  - Accurate to this modern understanding of soil is Nikiforoff's (1959) definition of soil as the "excited skin of the sub aerial part of the earth's crust".
  - Generally soil is the porous, powdery and unconsolidated outer layer of the Earth's crust which is formed by weathering of minerals and decomposition of organic substances. Soil is dynamic, three-dimensional (having length, width and depth) piece of landscape with a three phase (solid, liquid and gaseous) system. Soil is the basis and stratum of life hence some one rightly abbreviated the term 'Soil' as S = Stratum/Soul, O = Of, I = Infinite and L = Life.

- Complexity in soil genesis is more common than simplicity.
- Soil lie at the interface of Earth's atmosphere, biosphere, hydrosphere and lithosphere. Therefore, a thorough understanding of soil requires some knowledge of meteorology, climatology, ecology, biology, hydrology, geomorphology, geology and many other Earth sciences and natural sciences.
- Contemporary soil carry imprints of pedogenic processes that were active in the past, although in many cases these imprints are difficult to observe or quantify. Thus, knowledge of paleoecology, palaeogeography, glacial geology and paleoclimatology is important for the recognition and understanding of soil genesis and constitute a basis for predicting the future soil changes.
- Five major, external factors of soil formation (climate, organisms, relief, parent material and time), and several smaller, less identifiable ones, drive pedogenic processes and create soil patterns.
- Characteristics of soil and soil landscapes, e.g., the number, sizes, shapes and arrangements of soil bodies, each of which is characterized on the basis of soil horizons, degree of internal homogeneity, slope, aspect, landscape position, age and other properties and relationships, can be observed and measured.
- Distinctive bioclimatic regimes or combinations of pedogenic processes produce distinctive soils. Thus, distinctive, observable morphological features, e.g., illuvial clay accumulation in B horizons, are produced by certain combinations of pedogenic processes operative over varying periods of time.
- Pedogenic (soil-forming) processes act to both create and destroy order (anisotropy) within soils; these processes can proceed simultaneously. The resulting soil profile reflects the balance of these processes, present and past.
- The geological Principle of Uniformitarianism applies to soil, i.e., pedogenic processes active in soil today have been operating for long periods of time, back to the time of appearance of organisms on the land surface. These processes do, however, have varying degrees of expression and intensity over space and time.
- A succession of different soils may have developed, eroded and/or regressed at any particular site, as soil genetic factors and site factors, e.g., vegetation, sedimentation, geomorphology and change.
- There are very few old soils (in a geological sense) because they can be destroyed or buried by geological events, or modified by shifts in climate by virtue of their vulnerable position at the surface of the earth. Little of the soil continuum dates back beyond the Tertiary period and most soils and land surfaces are no older than the Pleistocene Epoch. However, preserved/lithified soils (paleosols) are an almost ubiquitous feature in terrestrial (land-based) environments throughout most of geologic time. Since they record evidence of ancient climate change, they present immense utility in understanding climate evolution throughout geologic history.

- Knowledge and understanding of the genesis of a soil is important in its classification and mapping.
- Soil classification systems cannot be based entirely on perceptions of genesis, however, because genetic processes are seldom observed and because pedogenic processes change over time.
- Knowledge of soil genesis is imperative and basic to soil use and management. Human influence on, or adjustment to, the factors and processes of soil formation can be best controlled and planned using knowledge about soil genesis.
- Soils are natural clay factories (clay includes both clay mineral structures and particles less than 2 μm in diameter). Shales worldwide are, to a considerable extent, simply soil clays that have been formed in the pedosphere and eroded and deposited in the ocean basins, to become lithified at a later date.

2. SOILS

Soil is the uppermost layer of Earth's crust which comprises fine particles, dead remains of organisms, various minerals and water. Soil is a dispersed polyphase system consisting of solid, liquid and gaseous phase. Each phase is physically or chemically different and mechanically separable. Presence of more than one phase, the soil is called heterogeneous system.

Functions of soil: There are four main functions of soil as discussed below:

1. Soil is the production of biomass, ensuring food, fodder, renewable energy and raw materials. This functions is the basis of human and animal life.
2. Soil is a filler and buffer and transformation agent between the atmosphere, the ground water and the plant cover is strongly influencing the water cycle at the Earth surface as well as the gas exchange between terrestrial and atmospheric systems, protecting the environment, including human begins against the contamination of the ground water and the food chain. Soils are the central link in the biotransformation of organic carbon and continually play a role in releasing CO_2 and other trace gases in to the atmosphere.
3. Soil serves as biological habitat and gene reserve, with a large variety of organism in and above the soil.
4. Soil is a memory, and therefore a geogenic and cultural heritage, forming an essential part of the landscape in which we live.

Components of soil: There are four major components of soil as discused below:

- **Solid phase:** Consists of soil solids or particles of minerals and organic origin. It forms the skeleton of the soil system.
- **Liquid phase:** Consists of soil water or soil solution filling the soil voids.
- **Gasseous phase:** Consists of gases or air occupying the void space not occupied by water.
- **Soil minerals:** The soil minerals can be classified into:

(i) Primary minerals
(ii) Secondary minerals
(iii) Accessory minerals

(i) Primary minerals: The soil and silt fractions of the soil solids are mainly made up of the primary minerals. Quartz and feldspar are important primary minerals. Quartz occurs in most of the soils and makes up about 50 to 90 per cent of sand and silt fractions. The quartz crystal consists of silica tetrahedral in which all four oxygen atoms are shared and there are no cleavage planes. Feldspar are alluminosilicates of potassium, sodium and calcium. The structure of feldspar consists of silica tetrahydra linked together by sharing each oxygen atom between adjacent tetrahydra resulting into a three dimentional network or tekto-silicate structure. Potassium feldspars are orthoclase, microline, adularia etc. These are common minerals of granite and gneiss rocks. The plagioclase feldspars series consists of albite, oligoclase, andesine etc.

(ii) Secondary minerals: Secondary minerals generally occur in the clay fraction of the soil. These minerals are also known as layer silicate minerals, phyllo-silicates or clay minerals. These minerals are primarily responsible for physical properties of soil like soil consistency, plasticity, swelling and physico-chemical properties like cation-exchange, ion fixation and release.

(iii) Accessory minerals: Accessory minerals occur in the sand and silt fractions in appreciable amounts and include pyroxenes and *amphiboles*, *olivines*, *apatite*, *Tourmaline*, *Zireon*, *Magnetite*, *Ilmanite*, *Rutile*, *Kyanite*, *Staurolite* etc. Sometimes, carbonate (Dolomite) and sulphur (Gypsum) bearing minerals also occur in soils.

Clay minerals: Clays are composed of extremely small crystalline particles of one or more members of a small group of minerals known as clay minerals. The clay minerals are essentially hydrous aluminum silicates with isomorphous replacements of aluminium with magnesium and iron with alkalies and alkaline. Earths present as essential constituents. Clay minerals also sometimes contain primary and accessory minerals and organic matter. Each clay particle is negatively charged, micelle which attracts thousands of positively charged ions such as K^+, Na^+, NH_4^+, Ca^{2+}, Mg^{2+} and H^+ as on ionic double layer. The charge of clays comes from ionizable hydrogen or hydroxyl ions, present on its surface e.g. (–Al–OH) or (–Si–OH) leaving temporarily negative charge on the oxygen as (–Al–O–) or (–Si–O–). These are two types of clays.

A. Silicate clay

B. Sesquioxidic clay

A. Silicate clay: These clay minerals are:

1. **Montmorillonite:** The minerals of montmorillonite or smectite isomorphous series are freely expandable layer silicate e.g. beidellite, montronite and saponite. These minerals occur as extremely small particles ranging in diameter from 0.01 to 1 μ. The spacing of the layer ranges from 12 to 18 Å and is variable with exchangeable cation species and degree of hydration.

Complete drying leads to an spacing of less than 10 Å. Full hydration can float the layers apart independent of each other.

The general chemical formula is (Na Ca)0.33(Al, Mg)$_2$Si$_4$O$_{10}$(OH)$_2$.n H$_2$O

This lattice is negatively charged and has adsorption capacity for cations 10–12 times more than that of kaolinite clays. These clays are developed in arid regions, poorly drained and more alkaline parent rocks.

2. **Micas or Illite:** Micas are common occurrence in the soils often associated with smectite or montmorillonite clays. The idealized end member of micas is dioctahedral muscovite and trioctahedral biotite. The chemical formula of muscovite is $K_{0.95}Fe_{0.08}Al_{2.91}Si_{3.04}O_{10}(OH)_2$ and the biotite is $K_{0.91}Fe_{2.42}Al_{1.83}Si_{2.69}O_{10}(OH)_2$.

   The structure of these clays is 2:1 silica and all alumina type similar to montmorillonite except they have potassium ion holding adjacent layers, together, so tightly that water cannot penetrate between the layers freely.

3. **Kaolinite:** One of the most extremely distributed clay mineral, having one sheet of silicon, tetrahedral, per sheet of alumina octahedral per layer. Thus it is 1:1 type clay has no substitution of Al^{+++} for Si^{+++} or Mg^{++} for Al^{+++}, so the net negative charge is less. Each layer has one plane of oxygen (O^{--}) replaced by hydroxyl (OH^-), which results in slurry hydrogen (-H-) bonds to the oxygen plane of adjacent layer, hence have very strong bonding and does not allow water to penetrate between the layers. Due to this property these clays do not swell in water and remain very tough. These clays are formed in humid, warm and well drained tropics and subtropics. The chemical formula of kaolinite is $Al_4[(OH)_8, Si_4, O_{10}]$.

   Kaolinite is a two-layer silicate mineral commonly occurring in soil.

B. Sesquioxide clays: In warm-humid climate the soils with persistent extension leaching, the weathered minerals e.g. soluble silica and much of alumina leach away. The remaining material having lower solubilities are sesquioxides. These are mixture of aluminium hydroxide ($Al(OH)_3$) and iron oxide (Fe_2O_3) or iron hydroxide ($Fe(OH)_3$) generally expressed as R_2O_3 where R is iron and aluminium. Due to presence of iron and aluminium, the colour of soil is red to yellow. These are intensively weathered soils for perhaps hundreds of thousands years.

Soil with high silica/sesquioxide ratio are called siliceous and those with low ratio are called sesquioxide soils. The sesquioxide clays do not swell, are least sticky and have quite different proportion for those of silicate clays. The high percentage of iron and aluminium hydrogen oxides furnishes and enormous surface for adsorption of phosphorus often lowers the availability of phosphorus.

Ralph E. Grim (1953) suggested the classification of clay minerals as shown below:

Classification of clay minerals

I. Amorphous

Allophane group

II. Crystalline

A. Two layer type (Sheet structure composed of units of one layer of silica tetrahedrons and one layer of alumina octahedrons)

1. Equidimensional

   Kaolinie group

   Kaolinite, nacrite etc.

2. Elongate

   Halloysite group

B. Three layer type (Sheet structure as composed of layers of silica tetrahedrons and one central dioctahedral or trioctahedral layer)

1. Expanding lattice

   a. Equidementional

   Montmorillonite group – montmorillonite, sauconite etc.

   Vermiculite

   b. Elongate

   - Montmorillonite group – montmorillonite, saponite, hectorite

2. Non expanding lattice

   Illite group

C. Regular mixed layer type (Ordered sticking of alternate layers of different type).

Chlorite group

D. Chain structure type (Horneblende like chains of silic tetrahedrons linked together by octahedral groups of oxygen and hydroxyls containing Al and Mg atoms)

- Atlapulgite
- Sepiolite
- Polygorskite

Vermiculite: Vermiculite also occurs extremely in soils and is a produce of hydrothermal alteration of mica. The structure of vermiculite consists of sheets of trioctahedral mica separated by layers of two water molecules thickners (4.98 Å).

Chlorite: The substitution of a complete charged hydroxyl sheet in 2:1 layers silicates gives a 2:1:1 layer or 2:2 layered structure of chlorite. Chlorite occurs extensively in soils and is a 2:2 layer silicate. A clay sample consisting predominantly of 2:1 type mineral is frequently mixed with 2:2 type of mineral resulting in mixed layering such as chlorite with vermiculite or chlorite with mica. The chemical formula is $Fe_{4.54}Al_{2.92}Si_{2.54}O_{10}(OH)_8$.

As the inter-layer potassium is subjected to depletion through chemical equilibrium with the soil solution, the exchangeable ions and inter-layer water is gained and leads to another mineral according to following relation:

Micas → Illite → Vermiculite → Montmorillonite

Beidellite: Substitution of Al^{3+} for Si^{4+} in silica sheet (tetrahedral).

Nontronite: Fe^{3+} (trivalent iron) dominates the alumina sheet (octahedral) and some Al^{3+} replaces Si^{4+} in silica sheet.

Cation exchange capacity of different clay minerals

Clay minerals	Cation exchange capacity (cmol(P^+)kg^{-1}
Kaolinite	3–10
Illite	10–40
Chlorite	10–40
Montmorillonite	80–150
Vermiculite	100–150
Organic colloids	>200

Silica-Sesquioxide ratio: The proportion of silica to sesquioxides in the clay fraction of soil expressed as a molar ratio. It may be about 6 in the original rock and ranges from 1.25 to 3.6 in most soils. Red and brown soils have 2.0-2.5 indicating the presence of free iron oxide and alumina by the leaching of silicic acid during weathering.

- 3.0 – 4.3 silica-sesquioxide ratio in black soil
- >2.50 silica-sesquioxide ratio in ash grey and black soil
- 2.0 – 2.5 silica-sesquioxide ratio in red and brown soil
- <1.33 silica-sesquioxide ratio in laterite soil
- The ratio is high in heavy (clayey) soils and low in coarse light soil.
- A high ratio means high base exchange, hygroscopic and field moisture capacity and fertility status.
- **Allopane (μ):** The most significant amorphous silicate mineral, poorly defined aluminium(μ) silicate, most prevalent in soils developed from volcanic ash and high capacity to adsorb cations.
- Silica dominant sheet is called tetrahedal because of its four sided configuration and Al and/or Mg sheet is called octahedral.
- Al- dominated sheet is called Di-octahedral sheet while Mg- dominated sheet is called trioctahedral.
- The tetrahedral and octahedral sheets are the fundamental structural units of silicate clays.

3. SOIL FORMATION

A graded change of consolidated mass (rock), not capable of growing plants to the unconsolidated (loose) layer of material that can support plant, if climate is suitable is called soil formation. The loose material after various weathering processes develops a soil profile, with different horizons and addition of organic material etc. is called soil development. Soil formation is the series of weathering and changes within a soil mass which is caused by many other factors simultaneously such as additions, losses, transportations and translocations.

Parent materials

The parent material is formed from original rock by the process of weathering and then it forms the soil. The soil forming minerals are classified into two categories:

1. **Primary minerals:** These are principles constituents of rocks e.g. Quartz, Feldspar, Mica, Calcite, Hornblende, Biotite, Augite, and Olivine etc.
2. **Secondary minerals:** There are the secondary products of weathering and various natural changes that take place in primary minerals e.g. Dolomite, Gypsum, Apatite, Hematite, Limonite, and clay minerals etc.

Weathering of rocks

The minerals are inorganic substances, in rocks from which soil is formed by the process of weathering.

Weathering is the process whereby the solid rocks of the earths crust are broken down to form the parent material of soil. There are mainly three types of weathering processes which occur in the soil.

A. Physical weathering: Physical weathering is the mechanical disintegration of the exposed surface of rocks accomplished by following processes.

1. **Expansion:** Contraction processes of non-conductor rocks, loosens the rocks to crumble by heat and cold.
2. **Freezing:** Creates massive force which enlarges the crevice, breaking off fragments or displacing masses. Thawing results rocks fragments to roll down the slopes.
3. **Grinding and Rubbing** of moving rocks against each other by water, wind, glacier result in very effective disintegration.

Physical weathering is also process which causes to change only in size but hastens the chemical action by exposing more reacting surfaces.

B. Chemical weathering: The chemical weathering caused the complex minerals to dissolve and make changes in their structure which is fragmented easily. In humid climates the chemical and physical processes occur simultaneously and markedly accelerate each other. The following chemical changes take place.

1. **Oxidation:** The rocks containing sulphides, carbonates and silicates of iron undergo oxidation as

$4FeCO_3 + O_2 \longrightarrow 2Fe_2CO_3 + 4CO_2$

$3MgFeSiO_4 + 4H_2O \longrightarrow H_2MgSi_2O_9 + SiO_2 + 3FeO$

Olivine Serpentine Silica Ferrous oxide

$4FeO + O_2 \longrightarrow 2Fe_2O_3 \longrightarrow 2Fe_2O_3.3H_2O$

Ferric oxide Limonite

The oxidized minerals weaken the rocks and allow them to crumble easily.

2. **Hydration:** Feldspar, amphiboles, pyroxene minerals rapidly associate with water and become soft by losing their luster and elasticity.

$2Fe_2O_3 + 3H_2O \longrightarrow 2Fe_2O_3.3H_2O$

$KAl.Si_3O_8 + H_2O \longrightarrow H.Al.Si_3O_8 + KOH$

Orthoclase Acid silicate (Soluble)

3. **Hydrolysis:** In the process of hydrolysis minerals split by water and form hydroxide of new more soluble substances. During hydrolysis processes the desilicification, dealkalization also occur simultaneously with the formation of new compounds.

$K_2O . Al_2O_3 . 6SiO_2 + 2H_2O \longrightarrow H_2O . Al_2O_3 . 6SiO_2 + 2KOH$

Feldspar

$H_2O . Al_2O_3 . 6SiO_2 + 2H_2O \longrightarrow Al_2O_3 . 2SiO_2 + 2H_2O + 6SiO_2$

Kaolinite Silica

$2H . AlSi_3O_8 + 8H_2O \longrightarrow Al_2O_33H_2O + 6SiO_2 . H_2O$

Acid aluminosilicate Hydrated alumina Hydrated silica (Soluble)

4. **Carbonation:** Decay of organic matter produces carbonic acid which dissolves minerals more rapidly. The acid readily reacts with sodalime, feldspar, pyroxene, amphiboles, causing the loss of bases from these minerals.

$CaCO_3 + H_2CO_3 \longrightarrow Ca(HCO_3)_2$

Calcite Carbonic acid (Soluble)

$Ca(Feldspar) + 2H_2CO_3 \longrightarrow H\text{-}(Silicate) + Ca(HCO_3)_2$

Carbonic acid (Soluble)

$K . Al . Si_3O_8 + 2H_2CO_3 \longrightarrow H_4Al_2Si_2O_9 + K_2CO_3 + 4 SiO_2$

Orthoclass Carbonic acid Aluminium silicate Silica (Soluble)

C. Biological weathering: The biological agents such as vegetation, human and animal actions accelerate the physical as well as chemical weathering. Plant roots help in widening the cracks and crevices in the rocks. Forming, mining, constructions and other human actions cause the quicker weathering process. Burrowing animals are important in biological weathering by exposing subsoil to weathering. The soil microbes continuously weaken the rocks by their biochemical process and enzymatic actions.

Differences between different kinds of wathering

Physical weathering/ Mechanical weathering	Chemical weathering	Biological weathering
It means disintegration of rocks and minerals	It means decomposition	It means disintegration and decomposition of rocks and minerals by biological or living agents
The primary minerals and rocks are splitted in fragment	Chemical weathering is not possible without the presence of water	Biological agent are men and animals, higher plants and their roots, microorganisms
Mechanical action of various weathering agents i.e. water, wind, ice, hail storm, frost, temperature and glacier	Weathering takes place mainly at the surface of rock minerals and formation of secondary products	
Consolidated massive rocks converted into smaller bites without any chemical change	Chemical agents help in weathering to some extent through the processes such as solution, hydration, hydrolysis, oxidation, reduction, carbonation etc.	
In physical weathering, size and shape of rocks and minerals may be changed		

Soil forming processes

A set of reactions leading to the formation of a group of soil is called soil forming process. The important soil formation processes are described below:

- **Calcification:** Under restricted rainfall, percolation of water through profile is not sufficient to remove carbonates, hence carbonates of Ca and Mg are redistributed in the profile and accumulated at percolated depths.
- **Podzolization:** Under humid conditions, vegetation and forests predominates which leave much of organic material deposits at the surface. The organic material decomposes thus organic acid produced, clay and iron removed from upper layer. The translocated materials are fractionated at different horizons of soil.
- **Laterization:** The hydrolysis of rock mineral produces iron and aluminium oxides, silicic acid and kaolin. The soluble nutrients are lost from soil. This process occurs mainly in tropical conditions with high rainfall and drought.
- **Salination and alkalization:** The process of accumulation of salts such as chlorides, sulphates, bicarbonates of sodium, calcium and magnesium called salination. On rainfall salts are removed (desalination) to result of deflocculation

of colloid with Na^+ ion adsorbed on colloidal complex. The soil solution becomes alkaline, the process is called alkalization.

- **Peat formation:** When water stands on the surface, the regular addition of organic matter build up in large amount under decomposed or partly decomposed matter. This process gives peaty soil, which occurs in forest of high rainfall areas. Solubility of Fe, Al, Mg, Ca increases in such organic soils.
- **Eluviation:** Leaching or removing from upper horizon to lower horizon is known as eluviation. When salts, elements, organic matter, small particles etc. are dissolved or coincide in water or suspension with water and goes down in lower horizons by percolation from the upper surface, the process is called eluviation and the upper horizons are called elluvial horizon. Generally 'A' horizon is called eluvial horizon.
- **Illuviation:** When precipitation and deposition of salts, elements, OM, small particles etc. occur in lower surface, the process is called illuviation and lower horizons are called alluvial horizon. Generally, 'B' horizon is also called illuvial horizon.
- **Cumulization:** Cumulization is the process of accumulation of mineral material on the soil surface either by air or water. Process occurs in where material eroded from upper slope has been accumulated in lower slope resulting heap in the lower portion.
- **Lessivage/clay migration:** Removal of fine clay in suspension form from the upper part of soil profile and its accumulation in the lower part is called clay migration. Physical downward movement of clay mineral is termed as clay migration.
- **Resilication.** Formation of Kaolinite from Gibbsite or formation of Smectite from Kaolinite in presence of large amount of $Si(OH)_4$ (silicic acid) at higher pH values.
- **Desication:** The chemical migration of silica out of a section of soil profile.
- **Humification:** The transfer of raw organic material into humus. Humification is a complex process involving various organisms i.e. fungi, bacteria, actinomycetes, earthworms, termites. Humification helps in the formation of surface humus layer.
- **Gleization:** The reduction of iron under anaerobic waterlogged soil condition with the production of bluish-greenish grey colour with or without yellowish brown, brown to black mottles and ferric or magniferrous concretions is called gleization. Gleization occurs in flood plain areas.

The formation of soil is a diverse and complex process which is affected by the five factors (climate, parent material, topography, time and biosphere) interacting with each other. The soil forming factors completely define the soil system, which can be expressed in the following equation.

$$S = F\ (Cl, b, r, p, t)$$

Where,

S= Soil, F = Factor, Cl = Climate, B = Biosphere, R = Topography, P = Parent material, T = Time or age of soil

Soil profile: Soil is a examined downwards, different layers of soils are usually found such section is called profile and its individual layers are called horizons. The horizons above the parent material are collectively called solum.

Soil horizons: The soil layers are grouped under three heads i.e. A, B and C. The sub-division of these are called horizons. The 'A' group called eluvial (translocated from one place to other) which lies at the surface, a zone of maximum leaching. The B group called illuvial, which is a zone of deposits and accumulation of specially Fe and Al. The C group is called the zone of parent material or unweathered material. The A group is the biggest which is further divided into following horizons:

- **AOO horizon:** Horizon consists of loose, relatively much of undecomposed organic debris, usually absent in grassland but present in forest land specially in temperate region.
- **AO horizon:** A partially or fully humified organic horizon, it may be malted, fibrous or granular on grassland, it is usually absent. In forest land, this zone is further divided into F (Fermentation) and H (Humified) zones.
- **A1 horizon:** This horizon is dark coloured mineral horizon and contains a large amount of humifies organic matter, thoroughly mixed with inorganic layer.
- **E or A2 horizon:** This light coloured mineral horizon formed due to leaching and bleaching action (Eluviation) which is well developed in Podzols.
- **A3 horizon:** This transition layer is often present like A and B.
- **B horizon:** This is a zone of maximum accumulation (Illuviation) of materials such as Fe and Al oxides and silicate clays, a characteristic zone of precipitated humus carrying iron and aluminium. In humid regions, silicate clays tend to accumulate in this zone. It is further divided into B1 transition layer, B2 illuviation layer, B3 transition layer of B and C.
- **C horizon:** This zone consists of consolidated parent material from which the solum has been developed. In this zone of least weathering and accumulation of carbonates of Ca and Mg, cementation, sometimes higher bulk density.
- **O horizon:** Organic horizon above mineral soil, occurring commonly in forest areas. Such horizon is visible in virgin soil and absent in arable soils.

Stages of soil formation

There are five stages of soil formation (Edward Mohr and Van Baren, 1954):

- **Initial stage:** Un-weathered parent material
- **Juvenile stage:** Weathering starts, but much of the original material still un-weathered.
- **Verile stage:** Easily weatherable minerals fairly decomposed, clay content increased, and slowly weathered minerals still persist.

- **Senle stage:** Decomposition reaches at final stage, only most resistant minerals survive.
- **Final stage:** Soil development is completed.

4. NUTRIENT CYCLE

A nutrient cycle is the movement and exchange of organic and inorganic matter back into the production of living matter. The process is regulated by food web pathways that decompose matter into mineral nutrients. Nutrient cycles occur within ecosystems. Ecosystems are interconnected systems where matter and energy flows and are exchanged as organisms feed, digest and migrate.

Ecosystem recycles locally, converting mineral nutrients into the production of biomass, and on a larger scale they participate in a global system of inputs and outputs where matter is exchanged and transported through a larger system of biogeochemical cycles.

Biogeochemical cycle is the movement (or cycling) of matter through a system (atmosphere, lithosphere, hydrosphere, biosphere). Biogeochemical include carbon cycle, nitrogen cycle, sulfur cycle, phosphorus cycle, iron cycle, water cycle, oxygen cycle and others that continually recycle along with other mineral nutrients into productive ecological nutrition.

There are two types of nutrient cycling viz., geochemical nutrient cycling and biological nutrient cycling.

1. **Geochemical nutrient cycling:** It is an open system which concerns the import export relationship of nutrients into and out of the ecosystem. It includes nutrient imports from such sources as nitrogen and fertilization. The outputs include leaching and erosion losses on drainage water, volatile losses from fire and desertification and removal in harvests.
2. **Biological nutrient cycling:** It is the closed system which involves plant-soil exchange of nutrients. It involves the transfer of nutrients between forest floor soil, associated plants and animal communities. It may also inlude the internal transfer of nutrients among organs within trees. Biological nutrient cycling deals with the translocation of nutrients within standing tree mass and nutrient transfer between trees and soils.

Major Pathways in Nutrient Cycling

The essential nutrients play various important roles in the structure and metabolism of plants. They are continually cycled through the atmosphere – soil – plant system according to specific transformation and transport processes. Some of these processes occur in soils and make elements available to plant roots and soil organisms.

(i) **Carbon cycle:** The carbon cycle is a bio-geochemical cycle which exchanges carbon between various reservoirs or storage places within the earth system viz. the land, the oceans and the atmosphere.

(ii) **Nitrogen cycle:** It is the biogeochemical cycle that describes the transformations of nitrogen containing compounds in nature. It is the transformastion of nitrogen from atmosphere (air) to soil, plants, animal life and returns back to air or soil through decay or denitrification.

(iii) **Phosphorus cycle:** The phosphorus cycle describes the movement of phosphorus through the lithosphere, hydrosphere and biosphere. The atmosphere does not play a significant role, because phosphorus and phosphorus based compounds are usually solids at the typical ranges of temperature and pressure found on earth.

(iv) **Sulphur cycle:** It is the cyclic movement of sulphur in different chemical forms from the environment to organisms and then back to the environment from organisms. It includes mineralization of organic sulphur into inorganic forms such as hydrogen sulphide (H_2S), elemental sulphur, as well as sulphide minerals and then oxidation of hydrogen sulphide, sulphide and elemental sulphur (S) to sulphate (SO_4^{2}). Finally reduction of sulphate to sulphide and incorporation of sulphide into organic compounds occurs.

(v) **Iron cycle:** The iron cycle is the biogeochemical cycle of iron through landforms, atmosphere and oceans. The cycling of iron consists largely of oxidation-reduction (redox) reactions that reduce ferric ion (Fe^{3+}) to ferrous ion (Fe^{2+}) and similarly oxidizes Fe^{2+} to Fe^{3+}.

5. SOIL ORGANIC MATTER

Soil organic matter consists of residues such as fallen leaves, immature flower and pods, roots, weeds, animal residues and wastes, worms, bacteria, fungi, algae, actinomycetes etc. The most cultivated soil contains 1-5 per cent organic matter in top 15-20 cm layer of soil and is responsible for improving soil structure, pore space, water holding capacity, nutrient holding capacity, biotic environment and reducing soil erosion. The organic residues undergo rapid decomposition in soil by aid of soil flora and fauna to form a more resistant product called humus. Soil humus is the complex array of substances left after extensive chemical and biochemical breakdown of plants and animal residues. Due to its complexity, humus is divided into three fractions such as fulvic acid, humic acid and humin. In addition to those fractions soil humus consists of polyuronides, sugar amine, proteins, nucleic acid, phospholipids, cellulose, vitamins, lignin, fats, tannins, waxes, resins etc. The complex nitrogenous compounds are called ligno-protein and composed of nitrogen, sulphur, phosphorus, and carbon in considerable amounts.

The nitrogen content in organic material is composed by carbon : nitrogen ratio (C:N ratio). Micororganisms need much nitrogen for their growth which they obtain by fixing atmospheric nitrogen. As progressive decay of organic material continues, much of the carbon escapes into atmosphere as carbon dioxide which narrows the C:N ratio. The plant residues with C:N ratio of 20:1 narrows down by decomposing microorganisms from 8 : 1 to 15 : 1. The decomposition of organic matter mainly depends upon temperature, humidity, soil moisture, soil texture, soil pH, nutrient content, management practices etc.

Soil organic matter comprises an accumulation of:

- Partially disintegrated and decomposed plant and animal residues.
- Other organic compounds synthesized by the soil microbes upon decay

Organic matter content of a well drained mineral soil is low : 1-6% by weight in the top soil and even less in the subsoil.

Origin of organic matter

The primary source of organic matter mainly originated from the plant tissue. Leaves and roots of trees, shrubs, grasses and other plants etc. usually supply large quantities of organic materials to the soil.

Animals are generally considered secondary source of organic matter. As they break down the original plant tissues, they contribute waste products and leave their own bodies after death. Earthworms, centipedes, insects and ants, also play an important role in the turnover of plant residues.

Nature and composition of soil organic matter

The native fraction of soil organic matter is made up of a heterogeneous mixture of polymerized aromatic molecules, polysaccharides, bound amino acids, uronic acid polymers and various organic phosphorus compounds. Organic residues consist of both organic and inorganic fractions.

A. Organic

(I) Nitrogenous organic compounds

- Water insoluble – protein, peptides, peptones and others carrying sulphur
- Water soluble – nitrates, ammoniacal compounds etc.

(II) Non-nitrogenous organic compounds

- Carbohydrates – cellulose, hemi cellulose, starch, pectin, sugars etc.
- Ether soluble – fats, oils, waxes, resins, steroids etc.
- Lignin – tannin, organic acids, essential oils etc.

B. Inorganic (mineral constituents or ash): Ca, Mg, Na, K, Fe, Mn, Zn, Cu etc.

Decomposition of organic matter

The organic materials (plant and animal residues) when incorporated in the soil, may once be attacked by organism and converted (decomposed) into new substances.

(A) Decomposition of soluble substances

(i) Sugar and water soluble nitrogenous compounds are first to be decomposed as they offer a very readily available source of carbon, nitrogen and energy for the microorganisms. The reaction is as under:

$$\text{Sugar} + \text{Oxygen} \longrightarrow \text{Carbon dioxide} + \text{water}$$

(ii) Ammonification: Soluble nitrogenous compounds, viz, amino acids, amides, ammonium compounds, nitrates etc., are also attacked by the microorganisms. The transformation of organic nitrogenous compounds into ammonia is called ammonification. The ammonification occurs as a result of action of enzymes produced by microorganisms. Their action is chiefly hydrolytic and oxidative.

Organic nitrogen ⟶ NH_3 or Protein ⟶ Polypeptides ⟶ Amino acids ⟶ Ammonia or salts of NH_4^+

Organic nitrogen ⟶ NH_3

(iii) Nitrification: The process of conversion of ammonia to nitrite (NO_2) and then to nitrate (NO_3) is known as nitrification. The production of nitrate is more rapid than that of nitrite, while the formation of ammonia is the slowest process. Nitrification is an aerobic process involving the production of nitrate from ammonia salts. It is the work of autotrophic bacteria.

$$\underset{\text{Ammonia}}{NH_3} \xrightarrow{\text{Nitrosomonas}} \underset{\text{Nitrite}}{NO_2} \xrightarrow{\text{Nitrobacter}} \underset{\text{Nitrate}}{NO_3}$$

The process of conversion of NH_4 to NO_2^- by nitrosomonas bacteria and then to NO_3^- by nitrobacter bacteria is known as nitrification. This involves two oxidation processes. The ammonia produced by ammonification is an energy rich substrate for nitrosomonas bacteria. They oxidise it to nitrite:

$$2NH_4^+ + 3O_2 \longrightarrow 2NO_2^- + 2H_2O + 4H^+ + 66 \text{ K Cal energy}$$

This in turn provides a substrate for nitrobacter bacteria which oxidises the nitrite to nitrate:

$$2NO_2^- + O_2 \longrightarrow 2NO_3^- + 18 \text{ K Cal energy}$$

This is the only source of energy for these prokaryotes and they are chemoautotrophs.

(iv) Denitrification: The process which involves conversion of soil nitrate into gaseous nitrogen or nitrous oxide is called denitrification. Water-logging (e.g., Rice field) and high pH will increase nitrogen loss by denitrification.

$$\text{Nitrate} \xrightarrow{\text{Pseudomonas/Bacillus}} \text{Nitrogen gas}$$

The process of conversion of soil NO_3^- (nitrates) into gaseous nitrogen or nitrous oxide by Pseudomonas, Clostridium or Bacillus bacteria is called denitrification. Denitrification is also called nitrate respiration.

$$2NO_3^- \longrightarrow 3O_2 + N_2\uparrow \text{ providing up to 238kJ}$$

$$2NO_2^- \longrightarrow 2O_2 + N_2\uparrow$$

The liberated oxygen is used as an electron acceptor in the processes that oxidise organic molecules, such as glucose.

(B) Decomposition of insoluble substances

(i) Breakdown of protein: The conversion of proteins into amino acids, amines and urea by the action of heterotrophic bacteria (*Rhizobium*, *Azotobactor*, *Clostridium*, *Amonifers*, *Denitrifers*) or fungi is known as Aminization. Bacteria in alkaline soil and neutral soils while fungi dominate in acidic soils. Amine ($R\text{-}NH_4$) is formed by Aminization.

Proteins $\xrightarrow{\text{hydrolysis}}$ Peptones, Peptides, Proteoses ⟶ Amino acids ⟶ Ammonia

(ii) **Breakdown of cellulose:** Cellulose is converted to cellubiose by cellulase enzyme and then to glucose by cellobiase enzymes in the process of enzymatic hydrolysis. Glucose is oxidized to CO_2 in aerobic condition and converted to organic acids and alcohols under anaerobic condition. Decomposition of cellulose is quite rapid in well aerated soils. It is slower in acid soils than in neutral and alkaline soils.

$$\text{Cellulose} \xrightarrow[\text{Cellulase Oxidation}]{\text{hydrolysis}} \text{Cellobiose} \xrightarrow[\text{Cellobiase Oxidation}]{\text{hydrolysis}} \text{Glucose}$$

(iii) **Breakdown of hemicelluloses:** Hemicellulose (polysachharides) is converted to oligosachharides by hemicellulase enzyme and then to simple sugars in the process of hydrolysis. Simple sugars are oxidized to CO_2 in aerobic condition and converted to organic acids and alcohols under anaerobic condition.

$$\text{Hemicellulose} \xrightarrow{\text{hydrolysis}} \text{sugar + uronic acids}$$

$$\text{Sugar} \longrightarrow \text{Alcohol, } CO_2 \longrightarrow \text{Organic acids}$$

(iv) **Breakdown of starch:** Starch (glucose of polymer) is converted to maltose by amylase emzyme and then to glucose by maltase and α –glucosidase, in the process called hydrolysis.

$$\text{Starch} \xrightarrow{\text{hydrolysis}} \text{Maltose}$$

$$\text{Maltose} \longrightarrow \text{Glucose}$$

(C) **Decomposition of Ether-soluble substances**

$$\text{Fat} \xrightarrow[\text{Lipase}]{\text{enzyme}} \text{Glycerol + fatty acids}$$

$$\text{Glycerol} \longrightarrow \text{Organic acids}$$

(D) **Decomposition of lignin:** Lignin decomposes slowly and much slower than cellulose. Complete oxidation of lignin gives rise to CO_2 and water.

$$\text{Lignin} \xrightarrow{\text{Oxidation}} CO_2 + \text{Water}$$

(E) **Simple decomposition products**

- **Carbon:** CO_2, CO^{--}, HCO_3^-, elemental carbon
- **Nitrogen :** NH_4^+, NO_2^-, NO_3^-, gaseous nitrogen
- **Sulphur :** S, H_2S, SO_3^{--}, SO_4^{--}, CS_2
- **Mineralization of sulphur**

$$S + 2O_2 \xrightarrow{\text{Enzymic}} SO_4^{-2}$$

Mineralization of organic phosphorus

$$\text{Phosphorus} \longrightarrow H_2PO_4^- \longrightarrow HPO_4^{--}$$

Carbon : Nitrogen (C:N) ratio

The ratio of the weight of organic carbon to the weight of total nitrogen (N) in a soil is known as C : N ratio. The ratio of carbon to nitrogen in cultivated soil commonly ranges from 8 : 1 to 15 : 1. Humus contains 10 : 1 to 12 : 1. The importance of C : N ratio is in mineralization and conserve N and organic matter (OM) in soil. If C : N ratio of organic matter is >20, it will results more N mineralization and more gain of available N in soil. If C : N ratio of OM is 20-30, results in no gain of available N. If C : N ratio of OM is >30, resulting less N-mineralization and more N- immobilization and finally loss of available N in soil. If decomposition of manure is slow, then wider/higher C : N ratio is present. If decomposition of manure is fast, then lower/narrow C : N ratio is observed. Similarly, C : N ratio of sub-surface soil is narrow than surface soil. It is reported that C : N ratio of dry areas soils is narrower than humid areas soil. C : N ratio of mature residues i.e. legumes or non legumes is higher than young succulent materials. C : N ratio of warmer regions (arid climate) is lower than colder regions (humid climate). C : N ratio decreases due to loss of carbon and conservation of nitrogen. C : N ratio of OM with less carbohydrate is narrow/lower than the OM with higher carbohydrate. Generally C : N ratio of Indian soils is observed as 14 : 1. Lower C : N ratio means greater amount of mineralization of N to NH_4^+ and NO_3^- form.

C : N ratio of different organic matter and soils

Organic matter/soil	C:N ratio	Organic matter/soil	C:N ratio
Soil microbes	4-9 : 1	Saw dust	400-500 : 1
Humus	10-12:1	Cultivated/arable soil	8-15 : 1
Green legumes	12 : 1	Mineral soil, waterlogged soil	10-12 : 1
FYM	20 : 1	Organic matter	40 : 1
Green grasses	40 : 1	Tropical soils	10 : 1
Maize stalk, sugarcane bagashi	40-50 : 1	Berseem	21 : 1
Grain straw, dry grasses	80 : 1	Alfa-alfa, clover	20 :1
Rice stover	89 : 1	Bacteria	5 : 1
Wheat straw	124 : 1	Actinomycetes	6 : 1
Jowar stalk	93 : 1	Fungi	10 : 1

Humus: Humus is more decomposed, complex and resistant mixture than organic matter. Humus is fairly stable and brown to black in colour whereas organic matter is not stable and colour is variable. Humus is a colloidal substance modified from organic matter. Humus possesses a definite C : N ratio or organic matter is very wide (40 : 1 to 12 : 1).

Humus properties: Humus is the colloidal complex of stable organic compounds. Humus contains higher proportion of organic compounds that are resistant to decomposition. The soil humus level represents the underlying level of the soil carbon store. As the fresh organic matter is the source of humus, more the OM, the more the soil carbon level. The colour of humus is brown, dark brown to black and the C:N ratio of humus is 10-12:1. CEC of humus is 100-300 Cmol (P^+)/kg soil and the density is between 1.3 and 1.5 Mg/m^3. Humus absorbs 80-90 per cent water under saturated condition.

Classification of humus: According to Molder, humus is classified into three groups on the basis of colour and solubility in water and alkali solution viz.

(i) Soluble in water: Crenic, Apocrenic acid

(ii) Soluble in alkali: Almic (brown), humic acid (black)

(iii) Insoluble in alkali: Almin, Humin

According to Oden, humus is classified into four categories viz.

(i) Humus coal

(ii) Humic acid

(iii) Hematomelanic acid

(iv) Fulvic acid

Composition of humus

Material basis		Weight basis	
Lignin	55%	Carbon	50–60%
Amino acid	35%	Hydrogen	3–4%
Carbohydrate	15%	Oxygen	30–40%
Hemicellulose	7%	Nitrogen	4–5%
Resin, fat, wax	3%	Phosphorus	0.4–0.5%
Others	5%		

Beneficial effect of soil organic matter

Organic matter in the soil serves many beneficial effects. These are:

1. It provide essential plant nutrient such as macro and micro elements.
2. It improves physical, chemical and biological conditions of soil. It also improves the soil tilth, soil aeration and percolation.
3. It improves the water and nutrient holding capacity of the soil. It prevents losses of plant nutrients from leaching and by soil erosion.
4. The fresh organic matter acts as a source of food and energy to numerous soil biota which decompose these organic matter for completing their life cycle and produces simpler products.
5. The organic matter keeps the soil fertile.
6. It keeps the soil environment balanced so that natural processes are maintained.
7. It decreases wind and water erosion.
8. It makes phosphorus more available in acid soil.
9. It reduces alkalinity in the soil
10. It makes potassium more available.
11. It increases buffering capacity of the soil which checks rapid chemical changes in pH or soil reaction.
12. Humus provides a storehouse for the exchangeable and available cations.
13. Fresh organic matter supplies food for earthworms, ants and rodents.

14. Organic matter creates a granular condition of soil which maintains favourable condition of aeration and permeability.
15. Water holding capacity of soil is increased and surface runoff, erosion etc., are reduced as there is good infiltration due to the addition of organic matter.
16. Organic matter serves as a source of energy for the microbes and as a reservoir of nutrients that are essential for plant growth.

6. SOIL WATER

Water contained in soil is called soil moisture. There are different forms of water found in the soil.

1. **Gravitational water:** After the soil is completely saturated, the excess infiltration that starts moving downwards due to the force of gravity to finally join the groundwater reservoir is known as gravitational water.
    - Movement of gravitational water down the soil profile depends upon the permeability
    - Gravitational water, though useful to plants, does not normally remain in place for long.
    - If it remains in the soil for too long, serious damage may take place to most crops due to lack of air in the soil except rice.
    - It is also called free water.
    - Gravitational water is retained in the soil below 1/3 atm.
2. **Capillary water:** After the draining of gravitational water, a certain amount of water will remain held within the soil pores due to capillary forces. This water retained in the soil pores by the force of molecular attraction is termed as capillary water.
    - This water is held in the pore space as a continuous film and can be removed by applying force greater than the capillary force.
    - A major portion of this water is utilized by plants.
    - Capillary water behaves according to law of capillary.
    - Moisture retained in the soil is between 1/3 atm to 31 atm.
    - Water held at tension of 15-31 atm is known as inner capillary water. This water is hardly available to plants.
3. **Hygroscopic water:** Hygroscopic water is that part of the water which a dry soil will absorbed in order to reach a static state with the atmospheric water (vapour).
    - It is held by the force of molecular attraction and can be removed only by heating.
    - Hygroscopic water is not available to plants but some microbial activities takes place.
    - It is non liquid water and moves in vapour forms.

- Hygroscopic water is tightly bound by the soil particles at greater than 31 atm tension.

4. **Combined water:** Combined water is chemically composed water. It is held in the soil particles with the chemical forces. Combined water is not available to plant.
5. **Crystalline water :** Crystalline water is present in intra space of minerals. This water binds with minerals.
6. **Vapour water:** Vapour water is present in soil in the form of vapour specially at higher temperature. This water is not important for plant growth.

Biological classification of water

1. **Superfluous water**
   (a) Moisture tension less than or equal to 1/3 atm.
   (b) Such water held at field capacity but not beneficial for higher plants.
   (c) Harmful for crops if too much.
   (d) This water also called gravitational water.
2. **Available water:** Portion of water lying between field capacity (1/3 atm) and wilting coefficient (15 atm).
3. **Unavailable water:** Water held in soil at the permanent wilting point. It includes hygroscopic water which is removed too slowly by plants to prevent wilting.

Retention of water by soil

The water is held on the surface of the colloidal and other particles and in the pores. The water is retained in the soil in following ways:

1. Cohesion and adhesion forces: The two basic forces are responsible for water relention in the soil.

- The attraction of molecules for each other i.e., cohesion.
- The attraction of water molecules for the solid surface of soil i.e., adhesion. By adhesion, soil hold water molecules rigidly at their soil-water interfaces.

Together, these forces make it possible for the soil solids to retain water.

2. Surface tension: Water behaves as if its surface is covered with a stretched elastic membrane. At the surface, the attraction of the air for the water molecules is much less than that of water molecules for each other. Consequently, there is a net downward force on the surface molecules, resulting in sort of a compressed film at the surface. This phenomena is called surface tension.

3. Polarity or Dipole characters: The retention of water molecules on the surface of clay micelle is based on the dipole character of the molecule of water. The water molecules are held by electrostatic forces that exist on the surface of colloidal particles. The surface force is far greater than the force of gravity so water may remain attached to the soil particle. The water may remain attached to the soil particle or move downwords into the lower layers, depending on the magnitude of the resultant force.

Methods of expression suctions

The forces responsible for retention of water is the soil after drainage has stopped are due to surface tension and surface attraction and are called surface moisture tension. This refers to the energy concept in moisture relationship. The force with which water is held is also termed as suction.

There are two units to express differences in energy levels of soil water.

1. pF Scale: The free enery is measured in terms of the height of a column of water required to produce necessary suction or pressure differences at a particulate soil moisture level. The pF, therefore, represents the logarithm of the height (in cm) of a colum of water to give the necessary suction. If water is allowed to be with drawn from a wet soil under suction, the pressure below the layer of wet soil is less than above it. The differences between these two pressures determines the amount of water retained in the soil.

pF values related to pressure and height of column of water

Height of column of water (cm)	Pressure (atmosphere)	pF value
1	0.001	0
10	0.01	1
10^2	0.1	2
10^3	1.0	3
10^4	10.0	4
10^5	100.0	5
10^6	1000.0	6
10^7	10000.0	7

2. Atmospheres or Bars: Atmosphere is the average air pressure at sea level. If the suction is very low, as occurs in the case of a wet soil containing the maximum amount of water that it can hold, the pressure difference is of the order of about 0.01 atmosphere or 1.0 pF equivalent to a column of water 10 cm in height. Similarly, if the pressure difference is 0.1 atmosphere the pF will be 2.0. Soil moisture constants can be expressed in terms of pF values. A soil that is saturated with water has pFO, while an oven dry soil has a pF 7.0. The maximum water holding capacity occurs at pF 1.0 and field capacity at pF 3.2.

Soil-water movement

There are three types of soil-water movement occurs in the soil. (i) Saturated flow, (ii) Unsaturated flow, and (iii) Water vapour movement.

(i) Saturated Flow: Condition of soil, when all large and small pores, are filled with water is called saturated. Saturated flow takes place when the soil is saturated. The direction of flow is from a zone of higher moisture potential to a lower moisture potential. Generally, water percolates down (vertically) into the lower layers. But horizontal flow also occurs with very less rapidly in comparison to vertical flow. Horizontal movement is much more evident in the clay soil, whereas vertical movement is much more evident in sandy loam soil. From a practical point of view saturated flow is very important.

(ii) Unsaturated flow: Soil pores contain some air as well as water is called unsaturated soil. Under field condition most soil-water movement occurs where the soil pores are not completely saturated with water. The soil macropores are mostly filled with air, and the micropores with water and some air. Water movement under these conditions is very slow compared to that occurring when the soil is saturated. Movement will be from a zone of low suction (thick moisture film) to one of high suction (thin moisture film). The two forces responsible for this suction are the

(a) attraction of soil solids for water (adhesion)

(b) capillarity

The water at any point of the soil column is held by soil particles with a force which is equal to magnitude, by opposite in direction to that of gravity. This force is called soil suction or soil moisture tension. The movement of moisture takes place because of the differences in water potential. The portion of soil that is moist is at a low water potential, while that which is dry is at a high water potential. Moisture moves from high potential to low potential this water rises upward in the capillary tube.

(iii) Water vapour movement: There are two types of water vapour movement.

(a) Internal movement, the change from the liquid to the vapour state takes place within the soil, that is in the soil pores.

(b) External movement, the phenomenon occurs at the land surface and the resulting vapour is lost to the atmosphere by diffusion and convection (surface evaporation).

The diffusion of water vapour from one area to another in soils does occur. Water vapour moves for high vapour pressure (in moist soil) to low pressure (in dry soil).

Importance of soil water

Water is essential for all forms of life. Water plays a very significant role in soil-plant growth relationships. There are following functions of water in plant growth and crop production.

1. Water serves as a solvent and carrier of food nutrients for the growth of plants. Water itself acts as a nutrient also.
2. Water helps in good structure formation.
3. Water protects plant from adverse conditions like drought, frost etc.
4. Water also maintains the turgidity in plant.

5. Excess or deficit of water in the soil is a limiting factor in seed germination and plant growth.
6. The process of weathering and soil formation depends on water.
7. The soil water is a great regulator of physical, chemical and biological activities in the soil.
8. Water is essential for the process of photosynthesis.
9. Good filth can be secured at right stage of moisture content.
10. The microorganism inhabiting the soil also require water for their metabolic activities.

7. PHYSICAL PROPERTIES OF SOIL

The physical properties of soil are mainly the properties of its composed particles varying greatly in their sizes. The soil particles divided into various groups are called separates. Important physical properties of soils are soil texture, soil structure, density, porosity, consistence, colour and temperature.

1. Soil texture: The mineral component constitutes the largest volume of soil mass and plays a vital role in determining most of the physical properties of soil. The relative percentage of different soil separates are called soil texture. The common textural classes of soil are described below according to their particle diameter.

Soil separates	Particle diameter (mm)	No. of particles in 1 gm of soil
Fine gravel	2.0-1.0	90
Sand- Coarse	1.0-0.5	722
Medium	0.5-0.1	46213
Fine	0.1-0.05	722074
Silt	0.05-0.002	57×10^5
Clay	Below 0.002	90×10^9

Soil textural class: On the basis of fundamentals soil particles are grouped into following classes:

- **Sands:** The soil consisting of more than 70% of sand separates are called sands. In sands individual grains can be seen and felt readily. They have irregular shape, greater pore space, lower water retention, lesser absorption of nutrients and maximum downwards movement of water.
- **Sandy loam:** This soil consisting largely of sand has enough silt and clay present to give it a small amount of stability. Individual sand grains can be seen and felt readily. Squeezed when moist, it will form a cast that will not only hold its shape when the pressure is released but will withstand careful handling without breaking. The stability of moist cast differentiates this soil from sand.
- **Loam:** The soil consisting mixtures of separates without domination of any group is called loams. Loam is an even mixture of the different sizes of sand, silt

and clay. These soils have good physical condition. The pore space, leaching, waterholding capacity lies between sand and clay soils.

- **Silt loam:** Silt loam consists of a moderate amount of fine sand, a small amount of clay and a large quantity of silt particles. It lumps in a dry, undisturbed state and appears quite cloddy but lumps can be pulverized readily, the soil then feels soft and flowery. When wet, silt loam runs together and puddled. Either dry or moist casts can be handled freely without breaking.
- **Clay loam:** Clay loam is fine textured soil which breaks into clods or lumps that are hard when dry. When a ball of moist soil is pressed between the thumbs and finger, it will form a thin ribbon that will break readily, barely sustaining its own weight. The soil is plastic and will form a cast that will withstand considerable handling.
- **Clay:** Clay is also fine textured soil that breaks into very hard clods or lumps when dry, and is plastic and unusually sticky when wet. The soil consisting 40% or more of clay particles are classed as clay soil. These are less porous, more sticky, have much water retention capacity, absorb more nutrients, possess cohesive and adhesive property. They form crumbs and tillage becomes difficult.

2. Soil structure: Arrangement of soil particle are called soil structure. The soil structure is described by shape, size and grades of soil particles. When the soil particles do not aggregate and separated by beach other are called structureless but when aggregate and form cluster, gave granular, crumb, platy, spheroidal, blocky, subgranular, prismatic or columnar structure and named accordingly.

Two forms of structureless condition are recognized:

- **Single grain:** Particles are easily distinguished. Particle are unattached to each other and each particle acts individually. Movement of the air and capillary water is highest. Such structure is usually found in sandy soil and upper surface of podsols.
- **Massive:** Individual particle adhere closely to each other but the mass lacks planes of weakness. When clayey soils are ploughed under wet condition, its individual particles adhere closely to each other and soil becomes massive or puddle. Such puddle soil retains water due to lack of pore spaces and thus is suitable for rice.
  - Soil conditions and characteristics such as water movement, heat transfer, aeration, bulk density, amount and size of pore space, availability of plant nutrients and growth of microorganisms are much infilterated by soil structure.
  - The important physical changes imposed by the farmer i.e., ploughing, cultivating, drainage, liming and manuring.

Importance of soil structures

- Soil structure influences the amount and nature of porosity.
- Structure controls the amount of water and air present in the soil and their movement and circulation are also controlled by soil structure.

- It affects tillage practices.
- Structure controls runoff and erosion.
- Platy structure normally hinders free drainage whereas sphere like structure (granular and crumby) helps in drainage.
- Crumby and granular structure provides optimum filtration, WHC, aeration and drainage.
- It also provides good habitat for microorganisms and supply of nutrients.

3. Density of soil: Density of soil represents weight of soil per unit volume as gram per cubic meter. The density of soil represents mainly in two popular concepts:

- **Bulk density:** The oven dry weight of a unit volume of soil, which includes pore space, organic materials is called bulk density. Bulk density can be used to estimate the differences in compaction of a given soil. Soil with low bulk density have favourable physical conditions, however, with high bulk density exhibits poor physical conditions. Soil having bulk density between 1.4 and 1.6 g/cm^3 is considered to be good for plant growth. It is expressed in g/cc or mega gram per cubic metre (Mg/m^3) or (g/cm^3).

  It can be calculated by using following formula:

$$= \frac{\text{Weight of oven dry soil}}{\text{Volume of soil} \times \text{density of water or volume of soil core}}$$

- **Particle density:** The weight (mass) per unit volume of the solid particle is called particle density. It is expressed in g/cm^3 or g/cc. The particle density of mineral soil varies from 2.6 to 2.75 g/cc whereas, of organic matter varies from 1.2 to 1.7 g/cc. With increase in organic matter, the density decreases correspondingly. Large particle density represents large proportion of heavy minerals such as magnetite, limonite, haematite and zircon. It can be calculated by using following formula:

$$= \frac{\text{Weight of soil}}{\text{Weight of water displaced by soil}}$$

4. Soil porosity: The available air space between soil particles is called pore space. Sandy soils have large and continuous pores whereas clays have very small pores which transport water very slowly. Attraction forces in the soil retain water within the fine pores, which results in a water logged soil, the poor aeration. To the growing plants, pore sizes are of more important than total pore space.

- Pores less than about 0.06 mm in diameter is known as micropores.
- Pores more than about 0.06 mm in diameter is known as macropores.
- Cropping reduces macropores space for rearly movement of air by about one half.

5. Soil colour: Colour of soil may be inherited from its parent material. Red soil develops from parent material, granite and schist. Dark coloured due to humus, absorb more heat than light coloured soil. White colour is common due to salt or carbonate deposits.

- It is due to either mineral matter or organic matter and mostly to both.

- Organic matter imparts black to dark grey tinge, iron compounds for red, brown and yellow tinge and silica lime and other salts give light, white and grey tinges.
- Red colour is related to unhydrated ferric oxide although manganese dioxide and partially hydrated iron oxide may also contribute to it.
- Yellow colour indicates more moist soil than red colour.
- Soil colour directly influences soil temperature.

6. Soil temperature: Plant growth as well as chemical and biological activities in the soil are greatly influenced by the soil temperature. Nitrogen fixation from atmosphere through microorganism is the maximum between 0 and 90°F. Soil temperature also regulates air movement in soil to some extent. The amount of heat that enters the soil is controlled by climate, colour of soil, altitude, vegetation and cover of the land. Soil temperature provides suitable medium for optimum growth, activity of microorganisms, germination, nutrients movement and water absorption by plants.

7. Soil permeability: Permeability is the measure of the ease with which water moves through a porus medium. Following Darcy's law, the permeability of soil is a measured by its coefficient of permeability, k is expressed in cm/hr or m/day. Permeability of soil samples is measured in the laboratory by using a permeameter. Field peameability is obtained by conducting in-situ intilteration tests using an infiltrometer. Soil permeability up to a rate of 1.3-20.0 mm/hr is considered as slow, 20-50 mm/hr as moderate, 50-250 mm/hr as rapid and greater than 250 mm/hr as very rapid.

8. Colloid formation: When a small amount of dry soil is poured into a glass of water, it is found that a small fraction of the soil gets dissolved into the water and the heavier particles gradually settle at the bottom. A certain amount of soil particles, however, neither gets dissolved nor settle but remain suspended, giving the water a muddy colour. These suspended particles which remain in water in colloidal form are held in place by molecular attraction and do not settle under normal conditions. Soils with large amounts of fine particles are prone to large scale colloid formation. Colloid formation by a soil has great practical significance since it is one of the important factors which influence its susceptibility to water borne erosion and soil salinity. Prevention of soil loss in colloidal form, even from a well protected field, is a difficult task.

9. Soil consistence: Soil consistence refer to the resistance of soil materials to deformation or rupture. It depends upon the degree and kind of force which attracts one molecule to another. It varies with size, shape and arrangement of soil particles and the nature of the water films around it, nature and amount of inorganic and organic colloids. When water dries from soil particles, soils develop greater cohesive force and mechanical strength.

10. Soil air: Gaseous phase of the soil is called soil air. Under moist field conditions macropores constitute air space. Ordinarily, the occupation of nearly one-third of the pore space in the soil by air and two-third by water constitutes the most favourable condition for plant growth.

11. Soil moisture: Soil particles possess the power to adsorb moisture from atmosphere. The water adsorbed is known as hygroscopic water. Soil water is subject to certain physical forces as a result of which it is retained in soil. There are two forces acting (i) force of attraction/adsorption, (ii) force of gravity. Surface force is greater than gravity.

8. CHEMICAL PROPERTIES OF SOIL

The most important soil chemical properties are:

1. Soil pH
2. Soil electrical conductivity
3. Soil colloids
4. Anion adsorption
5. Cation Exchange Capacity
6. Anion Exchange Capacity
7. Organic matter
8. Plant nutrients

1. Soil pH

pH is defined as the negative logarithm of hydrogen ion concentration or simply the log of the reciprocal of the hydrogen ion concentration (Jackson, 1973)

$$pH = -\log (H^+) = \log 1/(H^+)$$

The acidity or alkalinity of a solution can be expressed on the scale of acidity and alkalinity. The scale of acidity or alkalinity is called pH scale.

pH values		Classification
< 6.5	:	Acidic
6.5-7.5	:	Neutral
> 7.5	:	Alkaline

2. Soil electrical conductivity

The total amount of soluble salts present in the soil and is expressed as millimhos/cm or dSm^{-1}. As the amount of the soluble salts in a solution increases, the electrical conductivity also increases. This electrical conductivity is measured in terms of the resistance offered to the flow of current using a conductivity bridge.

Classification

EC (dSm^{-1}) at 25°C	Salinity effect	Remarks
<1.5	Harmless	Salinity effect negligible
1.5 – 3.0	Critical	Limits yield of sensitive crops
>3.0	Injurious	Limits yield of many crops

3. Soil colloids

The colloidal state refers to a two phase system in which one material in a very finely divided state is dispered through second phase e.g. solid in liquid – clay in water (dispersion of clay in water). The clay fraction of the soil contains particles less than 0.002 mm in size. Particles less than 0.001 mm size possess colloidal properties and are known as soil colloids.

Types of soil colloids: There are four major types of colloids present in soil as discussed below:

1. Layer silicate clays
2. Iron and alluminum oxide clays (sesquioxide clays)
3. Allophane and associated amorphous clays
4. Humus

4. Cation Exchange Capacity (CEC)

The sum total of the exchangeable cations that a soil can adsorb. The higher the CEC of soil the more cations it can retain. Soil differs in their capacities to hold exchangeable Na^+ and other cations. The CEC depends on the amount and kinds of clay and organic matter present. A high clay soil can hold more exchangeable cations than a low-clay soil.

5. Anion Exchange Capacity

The basic principles of cation exchange apply as well to anion exchanger, except that the charges on the colloids are positive and the exchange is among negatively charged anions. A simple example of an anion exchange reactions is

Micelle NO_3^-	+	Cl	$\longrightarrow$	Micelle Cl	+	NO_3
Soil solid		In soil solution		Soil solid		In soil solution

6. Anion adsorption

Adsorption of negative ion (anion) e.g. Cl^-, NO_3^-, SO_4^{2-} and H_2PO_4 on positively charged sites of clay and organic matter is known as anion adsorption.

Sources of positive charges are:

1. Isomorphous substitution.
2. Surface and exposed broken bonds of clay lattice.
3. Complex alumimium and iron hydroxy ions in acid soils.
4. pH dependent charges are important for anion exchange of organic matter.

7. Organic matter

Soil organic matter consists of a whole series of products which range from undecayed plants and animals tissues to fairly amorphous brown to black material bearing no terrace of the anatomical structure of the material that is normally defined as soil humus.

Soil organic matter comprises an accumulation of:

- Partially disintegrated and decomposed plant and animal residues.
- Other organic compounds are synthesized by the soil microbes upon decay.

Organic matter content of a well drained mineral soil is low : 1-6 % by weight in the top soil and even less in the subsoil.

8. Plant nutrients

Mineral nutrient is an inorganic ion obtained from the soil and required for plant growth. Plants need 17th elements for their growth and completion of life cycle *viz.*, C, H, O, N, P, K, Ca, Mg, S, Fe, Mn, Cu, B, Mo, Cl, Ni and Na is the latest addition of essential nutrients. There are following criteria for essentiality of elements in plant nutrients:

- The plant must be unable to grow normally or complete its life cycle in the absence of the element.
- The element is specific and cannot be replaced by other element
- The element plays a direct role in metabolism.
- The deficiency symptoms of the element can be corrected or prevented by the application of that element only.

9. BIOLOGICAL PROPERTIES OF SOIL

Soil biology is very important and plays a vital role in decomposition and nutrient addition to soil. The presence of micro flora and fauna and macro organisms like earthworms, rodents, ants, snails etc. have significant effect on soil physical and chemical properties.

These biological organisms digest plant and animal residues incorporated in the soil, release carbon-di-oxide into the atmosphere for recycling through higher plants and simultaneously creates excellent physical and chemical soil conditions.

Soil organism can be classified into two major groups i.e. plant and animal.

Important functions of soil microorganisms

The function or role of soil microorganism can be divided into (i) beneficial function and (ii) harmful function.

(i) Beneficial functions: The beneficial function of soil microorganism are:

- Change and decomposition in organic matter
- Fixation of atmospheric nitrogen
- Formation and development of soil

Changes and decomposition in organic matter

The organic materials when incorporated in the soil are attacked by the microorganisms. The microorganisms break up the various constituents of the organic materials and convert them into new substances.

Fixation of atmospheric nitrogen

The conversion of elemental nitrogen to readily available form by nitrogen fixing micro organisms is called biological nitrogen fixation. The nitrogen fixing micro organisms are : bacteria, algae and mycorrhizae (fungus root).

Formation and development of soil

1. **Decomposition of rocks and minerals:** The microorganisms help in weathering of rocks and formation of soil.
2. **Decomposition of organic matter:** They help in decomposition of organic matter into simple and available forms.
3. **Soil structure improvement:** Bacteria, fungi and algae aid in the development of desirable soil structure by their secretions of gummy substances. Addition of organic matter in the soil also helps in the development of good soil structure.
4. **Movement of soil material:** Organic acids and carbon dioxide that are released during decomposition make insoluble phosphates and other unavailable compounds more available to plants.

(ii) Harmful functions

1. Denitrification
2. Development of plant diseases
3. Formation of toxic compounds
4. Competition for nutrients

10. ION EXCHANGE

Ion exchange takes place collodis. It has great importance in agriculture.

- It influences the effect of fertilizer and fertilizer practices.
- It controls soil sturcutre and crumb formation.
- It is reponsible for imparting a stable structure.
- It controls the process of soil formation e.g. in the development of clay minerals.
- It influences the retention and libration of plant nutrients such as calcium, phosphores, potassium etc.
- It plays an important role in the reclamation of acid and alkali soils.

Ion exchange are of two types.

1. Cation exchange or base exchange
2. Inion exchange or acid exchange.

1. Cation exchange: The phenomenon of the exchange of cations between soil and salt solution is knwon as cation exchange or base exchange and the cations that takes place in the reaction are called exchangeable cations. Cation exchange reactions are reversible. In colodis, hydrogen ion is adsorbed more strongly than is the calcium (Ca^{++}). In the given reaction calcium and hydorgen is chemically equivalent:

$$Ca^{++}\boxed{\begin{array}{c}\text{Clay}\\ \text{micelle}\end{array}} + 2H^+ \rightleftharpoons \begin{array}{c}H^+\\ \\ H^+\end{array}\boxed{\begin{array}{c}\text{Clay}\\ \text{micelle}\end{array}} + Ca^{++}$$

In the following reaction when potassium chloride is applied in the soil, potassium pushes its way into the colloidal complex and forces out equivalent quantities calcium, hydrogen and other elements (e.g. M)

$$\begin{array}{c}Ca_{40}\\ Al_{20}\\ Ca_{40}\\ M_{20}\end{array}\boxed{\begin{array}{c}\text{Clay}\\ \text{micelle}\end{array}} + 7KCl \rightleftharpoons \begin{array}{c}K_7\\ Ca_{38}\\ Al_{20}\\ H_{39}\\ M_{18}\end{array}\boxed{\begin{array}{c}\text{Clay}\\ \text{micelle}\end{array}} + 2CaCl_2 + 2MCl + 3HCl$$

Cation Exchange Capacity (C.E.C.)

The capacity of the colloidal complex to exchange all its cations with the cations of electrolyte solution is known as cation exchange capacity of soil. It also represents the soil total cation adsorbing capacity of a soil. Cation exchange increases with pH. At a very low pH value, CEC is higher and at high pH, CEC is relatively lower.

Factors affecting the CEC: The following factors affect the CEC:

- Clay soils tends to have higher cation exchange capacity than sandy soils. CEC for clay soils usually exceeds 30 me/100 g while the value ranges from 0 to 5 for sandy soils.
- The CEC of most soils increases with pH.
- Higher organic matter content in a soil have higher CEC.
- Montmorillonite has higher CEC than illite or kaolinite clay.

Percentage base saturation of soils

Hydrogen and aluminium tends to dominate acid soils, both contributing to the concentration of H^+ ions in the soil solution. Absorbed hydrogen contributes directly to the H^+ ion concentration in the soil. Al^{+++} ions do so indirectly through hydrolysis.

$$Al^{+++} + H_2O \longrightarrow Al(OH)^{++} + H^+$$

$$A(OH)^{++} + H_2O \longrightarrow Al(OH)_2^{\ +} + H^+$$

Most of the cations (Ca^{++}, Mg^{++}) called exchangeable bases, neutralise soil acidity. The proportion of the cation exchange capacity occupied by these bases is called the percentage base saturation. A definite correlation exists between the percentage base saturation of a soil and its pH.

Role of cation exchange: It is very important chemical reaction for the availability of plant nutrients in the soil. In normal fertile soils the total exchangeable bases (Ca, Mg, K, Na) constitute about 80–90% of the cation adsorbing capacity. Exchangeable hydrogen is usually under 20%. In these soils pH varies from 6.5 to 7.5.

The predominance of exchangeable calcium give rise to clay which imparts neutral reaction. When the proportion of exchangeable H^+ is high it give rise to acid soil (H-clay). It is called base unsaturated soil. The pH is less than 7.0. When exchangeable sodium form more than 10–15% of the total exchangeable cations it gives rise to alkaline soils. The pH value is greater than 8.0.

Cation exchange and soil fortility: Cation exchange capacity is the best index of soil fortility. By cation exchanges, hydrogen ions from the root hairs and microorganisms replace nitrient cations from the exchange complex.

(i) Cation saturation and soil fertility: Soil with a high calcium base saturation are the most satisfactory physical and nitritional condition.

Base unsaturated soils are acidic in nature due to exchangeable hydrogen. These soils are less fertile. Base saturated soils with dominant sodium cations are alkaline is nature. Alkaline soils are not fertile due to deflocculation, stickiness, hard to work, poor drainage and poor aeration.

(ii) Cation exchange and soil fertility: Due to the property of cation exchange the soluble inorganic fertilizer nutrients are not washed away from the soil e.g., ammonium ion.

Ammonium ions are taken up by plants. This process checks nitrient losses by leaching and make the soil fertile. The cations Ca, Mg, K and NH_4 are held on the colloidal surfaces and are readily available to plants.

(iii) Influence of complementary adsorbed cations and soil fertility: The order of strength of adsorption, when the ions are present in equivalent quantities, is as below.

$$Al^{3+} > Ca^{2+} > Mg^{2+} > K^+ = NH_4^+ > Na^+$$

Consequently, a nutrient cation such as K^+ is less tightly held by the colloids if the complementary ions are Al^{3+} and H^+ (acid soils) than if they are $Mg^{++} + Na^+$ (neutral to alkaline soils). The loosely held K^+ ions are readily available for adsorption by plants or for leaching in acid soils.

2. Anion exchange: The processes of anion exchange is similar to that of cation exchange. Colloides becomes positively charged due to addition of H^+ in OH^- resulted in net positive charge (OH_2^+). The capacity of holding anions increases with the increase in acidity. H_2PO_4 is adsorbed slightly at low pH but none at neutral soil, while NO_3^- ions are not adsorbed at all. In general, the relative order of anions exchange is:

$$OH^- > H_2PO_4^- > SO_4^{--} > NO_3^-$$

Importance of anion exchange: The adsorption of phosphate ions by clay particles from soil solution reduces its availability to plants. The phenomenon is known as phosphate fixation. The adsorption of phosphate ions by clay particles from soil solution reduces its availability to plants.

$$\underset{\text{(Available)}}{OH\text{–}clay + H_2PO_4^-} \rightleftharpoons \underset{\text{(Unavailable)}}{H_2PO_4 - clay} + OH^-$$

The phasphate ions, react with the hydrous oxides and get fixed, forming insoluble hydrogen phosphates of iron and aluminium.

$$\underset{\text{(Available)}}{Al(OH)_3 + H_2PO_4} \rightleftharpoons \underset{\text{(Unavailable)}}{Al(OH_2) \cdot H_2PO_4} + OH^-$$

If this reaction takes place under conditions of slight acidity it is reversible, and soluble phosphate is again liberated when hydroxy-phosphate comes in contact with ions. It is reaction takes place at a low pH under strongly acid condition, the phosphate (ions) are irreversible fixed and the totally unavailable for the use of plants.

11. ELECTRICAL CONDUCTIVITY OF SOIL

Electrical Conductivity (EC) is the ability of a material to transmit (conduct) an electrical current and is commonly expressed in units of milli Siemens per meter (mS/m). Soil EC measurements may also be reported in units of deci Siemens per meter (dS/m), which is equal to the reading in mS/m divided by 100. These are the standard units of measure of bulk soil conductivity. A Siemen is a measurement of the material's conductance.

Soil electrical conductivity (CE) is a measure of the amount of salts in soil (salinity of soil). It is an important indicator of soil health. It affects crop yields, crop suitability, plant nutrient availability, an activity of soil microorganisms which influence key soil processes including the emission of green houses gases such as nitrogen oxides, methane and carbon dioxide. Excess salts hinder plant growth by affecting the soil water balance. Soil containing excess salts occur naturally in arid and semiarid climates. Salts levels can increases as a result of cropping, irrigation and land management. Although EC does not provide a direct measurement of specific ions or salt compounds, it has been correlated to concentration of nitrates, potassium, sodium, chloride, sulphate and ammonia. For certain non-saline soils, determining EC can be a convenient and economical way to estimate the amount of nitrogen available for plant growth.

Soil electrical conductivity (EC) is a measurement that correlates with soil properties that affect crop productivity including soil texture, cation exchange capacity (CEC), drainage conditions, organic matter levels, salinity and sub soil characteristics.

Factors affecting soil electrical conductivity

Factors affecting EC include soil minerals, climate and soil texture which can not be changed. Salts originate from disintegration (weathering) of minerals and rocks. In areas with high amounts of rain fall, soluble salts from minerals and rocks are flushed below the root zone, eventually into deep ground water systems or into streams that transport salts to the ocean. Conversely, in arid areas or in areas where less rainfall occurs or saline irrigation water is used, soluble salts are more likely to accumulate the near the soil surface, resulting in high EC. Salt affected soils mainly occur in arid or semi arid areas where annual rainfall in low. Because salts move with water, iow areas, depressions or other wet areas where water accumulates tends to be higher in EC than surrounding higher lying, better drained areas. Clay soils dominated by clay minerals that have a high cation exchange capacity (CEC), such as smeetite, have higher EC than clay soils dominated by clay minerals that have a low CEC, such as kadinite. Soil with restricted layers, such as clay pans, typically have higher EC because salts cannot be leached from the root zone and accumulate on the surface. Infiltration water can also interact with underlying bed rock that weathers releasing salts which creates saline seeps where it exists.

Soil EC is also affected by cropping, irrigation, land use, and application of fertilizer, manure and compost. When managing for salinity on irrigated land, irrigation water salinity must also be measured. Irrigating in amounts too low to leach salts or with water high in salts, allows salts to accumulate in the root zone thus increasing EC.

Existing salinity levels and amount of salt contained in manure and municipal waste need to be closely monitored to prevent salinity problems, especially in arid climate. Nitrogen fertilizer application can increase salinity and should be monitored closely, especially on sites with potential salinity concerns. Management that leads to low organic matter, poor infiltration, poor drainage, saturated soil, or compaction can increases EC and the soils ability to buffer EC.

When irrigating, additional water beyond crop needs can be used to flush excessive salts below the root zone and maintain a EC level, which is based on crop tolerance. Care is needed when applying extra water to leach salts because soils can become waterlogged, allowing salts to accumulate. Leaving crop residue on the surface limits evaporation, and retains soil moisture allowing rainfall and irrigation to be more effective in leaching salts. In some cases, a combination of irrigation, and drainage is necessary to reduce salinity. Good soil management measures to maintain soil organic matter and overall soil health, must be utilized to maintain the desired EC level.

Relationship of EC to soil function

Electrical conductivity levels can serve as an indirect indicator of the amount of water and water soluble nutrients available for plant uptake such as nitrate-N. Arears of saline soils need to be identified and managed differently from areas of non-saline soils. Soil microorganism activity dectine as EC increases. This impacts important soi' processes such as respiration, residue decomposition, nitrification and denitrification.

Soils with a high concentration of sodium salts (sodic conditions) have additional problems, such as poor soil structure, poor infiltration or drainage as poor soil structure, poor infiltration or drainage and toxicity for many crops. Each crop has a salt tolerance.

$EC_{1:1}$ readings less than 1dS/m, soil are considered non-saline and do not impact most crops and soil microbial processes. EC 1:1 reading greater than 1 dS/m, are considered saline and impact important microbial process, such as nitrogen cycling, production of nitrous and other N oxide, gases, respiration and decomposition, populations of plant-parasite numatodes can increase and increased nitrogen losses.

12. REDOX POTENTIAL

Redox potential is defined as the specific indicator of the extent to which the oxidizing as well as reducing powers of a substance which has both reducing and oxidizing ingredient have achieved equilibrium.

Redox potential is used as:

- An indicator of electro chemical reactivity of substances in environmental conditions.
- For predicting corrosion protection of various substances and systems.
- Redox is a shortened form of the term oxidation-reduction.
- Redox potential of a system can be considered as a measure of the intensity of its oxidizing or reducing power, depending on the electro chemical balance.
- Oxidation potential measures the power of a substance or system to add oxygen or to remove hydrogen as well as to lose electrons. Reduction potential indicates the power to add hydrogen, less oxygen or attract electrons.
- As the redox potential increases in value and turns positive, its ability to oxidize is enhanced. When it decreases in value and turns negative, its reducing ability is quantitatively enhanced. It has some resemblance to pH value of a liquid.
- Any redox reaction can be analyzed as two half-reactions: One half in which a material constituent is oxidized and the other half in which process, another constituent is reduced.
- For the half-reaction such as reduction, the driving power is the negative redox potential. For oxidation, the driving power is the positive redox potential.
- As corrosion involves both oxidation as well as reduce redox potential beccome an indicator of the possibility of electro chemical activity likely to lead to corrosion or resistance to corrosion.
- Redox potential is a measure of the propensity of a chemical or biological species to either acquire or lose electrons through ionization.
- A species with a higher reduction potential possesses a higher tendency to acquire electrons and be reduced. Conversely, a species with a higher oxidation potential possesses a higher tendency to lose electrons and be oxidized.
- The reduction and oxidation potentials for a given species are identical in value but opposite in sign.

Oxidation-reduction potential (ORP):

One of the characteristics of water containing of water containing dissolved molecular hydrogen (such as ionized water) is that it exists a negative oxidation-reduction potential (ORP). Chemical reactions occurring in an aqueous solution are called redox reaction. The ORP measures the capacity of a solution to either release or accept electrons from chemical reactions. The ORP value, much like pH, is important for determining water quality and for water treatment processes.

Just as pH determines the waters relative state in receiving or ganing protons (i.e., hydrogen ions, H^+), ORP indicates the water's relative state to receive or gain electrons. A solution with a higher (more positive) ORP has the potential to oxidize a solution with a lower ORP. For example, +700 mV is an antioxidant to +900 mV.

Oxidation and reduction

Oxidation occurs when electrons are removed, such as when a free radical steals an electron from a cell. The cell is oxidized (an increase in its oxidation state), whereas, the free radical is reduced.

Therefore reduction means a receiving or gaining of electrons (decrease in its oxidation state), such as when an oxidant donates an electron to a free radical. The free radical is reduced and the antioxidant is oxidized. The antioxident is said to be the reducing agent, it reduces the free radical. Whereas the free radical is called the "Oxidizing agent", it oxidizes the antioxidant.

Potential

The potential is a characteristics of the chemical species to undergo an oxidation-reduction reaction. It is not the actual reaction. Rather, it is stored energy that has the ability to do work and is measured in volts, thus, the greater the voltage potential, the greater the ability and propensity to undergo a redox reaction.

The higher the species positive potential (i.e., free radical) the greater affinity it has for electrons, and thus a higher tendency to oxidize (steal electrons) from another species, like a cell membrane. Conversely the lower the species redox potential (i.e., antioxidants) the lower its affinity for electrons, and thus a higher tendency to donate its electrons and be a reducing agent (neutralize a free radical).

- It is important to note that just because something has a high ORP (negative or positive) does not mean a reaction will take place. Just like the depiction with the ball at the top of the hill, it has the potential to go down the hill, but someone still needs to give it the initial push (this is called activation energy, Ea). If the required activation energy is too high, the reaction may never occur. In fact, although we hear that diamonds last forever, they actually have a favourable potential to spontaneously convert to graphite (Gibbs free energy = –2.90 kJ/mol), but the Ea is so high that the reaction does not readily occur.
- It is also important to note that just because something has a negative ORP does not mean that it has any physiological antioxidant value. It is the chemical species responsible for producing the ORP value that determines whether or not it can act as a biological antioxidant. For example, a negative ORP can be produced by the addition of vitamin C, molecular hydrogen or aluminium metal to water, but only vitamin C and molecular hydrogen have physiological benefits-even though small amounts of aluminium (Al) in different oxidation states (i.e., Al^{+} to Al^{3+}) can give an ORP of over –700 mV. In fact, the metal can actually act as an oxidant, causing the very thing you are trying to prevent. This is important to keep in mind with many commercial products now tauting the benefits of ORP water.

13. PLANT NUTRITION

Plants, human and animals require food for growth and development. This food is composed of certain chemical elements often referred to as plant nutrients or plant-food-elements. Some of the important elements essential for plants are also essential for human and animal beings.

1. **Elements essential for plant (17):** Carbon, Hydrogen, Oxygen, Nitrogen, Phosphorus, Potassium, Calcium, Magnesium, Sulphur, Iron, Manganese, Boron, Zinc, Copper, Molebdenum, Sodium and Chlorine.
2. **Elements essential for human (19):** Carbon, Hydrogen, Oxygen, Nitrogen, Phosphorus, Potassium, Calcium, Magnesium, Sulphur, Iron, Manganese, Zinc, Copper, Molebdenum, Chlorine, Sodium, Iodine, Cobalt and Fluorine.
3. **Elements essential for animals (18):** Carbon, Hydrogen, Oxygen, Nitrogen, Phosphorus, Potassium, Calcium, Magnesium, Sulphur, Iron, Manganese, Zinc, Copper, Molebdenum, Chlorine, Sodium, Iodine and Cobalt.

Essentiality of elements in plant nutrition: Plant absorbs large number of elements from the soil and other sources during its growth and development, but all the elements are not essential for plant growth. The element must fulfill the following three requirement for essentiality (Arnon and Stout, 1939) which was refined by Arnon (1954):

1. A deficiency of the element makes it impossible for the plant to complete the vegetative and reproductive stage of its life cycle.
2. The deficiency is specific to the element in question and as such can be prevented or corrected only supplying that particular nutrient (element) to the plant.
3. The element must have a direct influence on the plant and must be directly involved in the nutrition and metabolism of the plant quite apart from its possible effects in correcting some unfavourable microbiological or chemical condition of the soil or other growing medium.

Essential plant nutrients

Sources from air

1. Carbon

Source from water

2. Oxygen
3. Hydrogen

Source from Soil

4. Nitrogen
5. Phosphorus
6. Potassium
7. Calcium
8. Magnesium
9. Sulphur
10. Iron
11. Manganese
12. Boron
13. Zinc
14. Copper
15. Molybdenum
16. Chlorine

These are 16 essential plant nutrients, sodium is also considered as essential plant nutrients. Now-a-days 17 are known to be essential plant nutrients.

On the basis of requirement of these elements by plants are classified into following groups:

1. **Basic nutrients:** Most of plant tissues, in general from 94 to 99.5 per cent is made up of carbon, oxygen and hydrogen i.e. nutrients obtained from air and water and only 0.5 to perhaps 5 to 6 per cent of the plant tissue are synthesized from soil constituents.
2. **Macro elements:** Out of 17 essential plant nutrients, six elements are used by plants in large quantity. These are nitrogen, phosphorus, potassium, calcium, magnesium and sulphur. Since these nutrients are utilized in relatively large quantity, they are designated as macronutrients. Nitrogen, phosphorus and potassium are utilized by plants in considerably large quantity (more than 1 ppm) are called macro/major nutrients.

   The calcium, magnesium and sulphur are sometimes called secondary nutrients due to their secondary importance.
3. **Micro nutrients:** Eight nutrients, namely, iron, manganese, boron, zinc, copper, molybdenum, sodium and chlorine are used by the plants in very small quantity (less than 1 ppm), these are therefore called micro/trace/minor/rare/oligo/spurne elements. The micro nutrients are, however, just as essential to plant growth as the macro nutrients.
4. **Ultra nutrients:** The concentration of these nutrients is 1 ppb (part per billion). Nicholas (1963) put forward this concept. Ultra nutrients are Mo and Co. At present this concept is not accepted.
5. **Beneficial nutrients:** Beneficial nutrients is also called potential micronutrients. The requirement of such elements at very low concentrations and often under specific conditions have been shown to stimulate the growth of certain plants e.g. Selenium (Se), Al, Rb, Sr, Ni, Cromium (Cr) and Arsenic (As).

Non essential but useful plant nutrients:

Besides the seventeen essential nutrients, there are other minor elements such as Sodium, Cobalt, Iodine, Strontium. Fluorine, Vanadium and Bromine, which may increase crop growth when present in soluble form.

Forms in which nutrients are utilized by plants:

Plant absorb nutrients from the soil as

(i) Single nutrients, uncombined with other elements.

(ii) Essential nutrient, combined with other elements to form nitrate, phosphate, sulphate etc.

Potassium	K^{+}
Calcium	Ca^{++}
Magnesium	Mg^{++}

Iron	Fe^{++}(Ferous), Fe^{+++} (Feric)
Manganese	Mn^{++} (Magnous), Mn^{+++}(Manganic)
Copper	Cu^{+} (Cuprous), Cu^{++} (Cupric)
Zinc	Zn^{++}
Chlorine	Cl^{-}

Nutrients absorbed in combined form or as salts:

Nitrogen	NH_4^+ (Ammonia), NO_2^- (Nitrate)
Phosphorus	PO_4^{--}, HPO_4^{--}, $H_2PO_4^-$
Sulphur	SO_4^- (Sulphite), SO_4^{--} (Sulphate)
Boron	BO_3^{--} (Borate), $HB_4O_4^-$ (Bimovate)
Molybdenum	$HMoO_4^-$ (Molybdate)
Carbon	CO_3^{--} (Carbamate), HCO_3^- (Bicarbamate)
Water	H^+, OH^-

- Mineral nutrients are available to the plant both in ionic and molecular forms, proportions depending on the nature and conditions of the soil solutions.
- Many compounds or salts when in solution exist as electrically charged portions called ions, most of the nutrient uptake by plants is in the form of ions.
- The ions penetrate the absorbing tissues of plants with greater ease than the molecular forms or compounds or salts.
- Some nutrient elements occur in two or more different ionic forms.
- Nitrogen is absorbed in either the ammonical or nitrate form. The nitrate ion can be superior for use by most of plants under many varied soil conditions such as strongly alkaline, strongly acidic, waterlogged and dry region etc.
- Under rice cultivation ammonical ion has proved superior.
- Most plants prefer ammonical and nitrate forms equally in their early growth stages.
- In alkaline condition, the PO_4^{2-} is predominant, however, in neutral and slight to moderate acid conditions the $HPO_4^{--}$and $H_2PO_4^-$ ions prevail. Under strongly acid conditions, the HPO_4^{2-} is largely predominant.
- Sulphur is absorbed mostly in the form of SO_4^- ion under favaourable conditions of oxidation by micro organisms. The SO_2 (gas) molecule from the air can be utilized by green plants to some extent.
- Iron, manganese and copper are absorbed largely in divalent ionic form, while boron and molybdenum are absorbed in monovalent ionic forms.
- The predominance of any ionic form is largely governed by the oxidation-reduction conditions of the soil.

If soil is well aerated, the ion of higher valency in each case tends to predominate, however, in poorly drained soils, ions with lower ionic valency will be present.

Mobility of nutrients

In the soil:

- **Mobile nutrients** are highly soluble and are not absorbed by clay complex e.g. N, S, B, Mn and Cl.
- **Less mobile** elements are also soluble but absorbed on clay complex and thus their mobility is reduced e.g., N, K, Ca, Mg and Cu.
- **Immobile** elements are highly reactive and get fixed in the soil e.g. P and Zn.

In plants:

- **Highly mobile** e.g. N, P and K
- **Moderately mobile** e.g. Zn
- **Less mobile** e.g. S, Fe, Cu, Mn, Mo and Cl
- **Immobile** e.g. Ca and B

Manures: These are organic source materials which are prepared by available natural resources (plant biomass) and animal waste (dung and urine) and contain plant nutrients in small quantities. The availability of plant nutrients in very slow. These are farm yard manure, compost, meals and cakes etc.

Fertilizers: These are inorganic material which are manufactured in the factories and have one or two or three essential plant nutrients in large quantities. The availability of nutrients from these fertilizers is very quick. These are urea, SSP, DAP, and MOP etc.

Losses of plant nutrients from the soil:

1. Removal of nutrients by plants uptake
2. Removal of nutrients by weeds
3. Removal of nutrients by leaching
4. Removal of nutrients by erosion
5. Removal of nutrients in gaseous form

Sources of plant nutrients:

1. **Commercial fertilizers, inorganic and organic:** Inorganic commercial fertilizers are Ammonium sulphate, Calcium ammonium nitrate, Super phosphate, and Muriate of potash etc.

   Organic commercial fertilizers are urea and Calcium cyanamide
2. **Bulky organic manures and concentrated organic manures** such as farm yard manure (FYM), night soil, compost, sludge, oil cakes, and blood meal etc.
3. **Green manure crops:** When legumes such as sunhemp, dhaincha, mung, urd etc are used as green manure crops, they added as much as 60 to 80 kg N/ha.
4. **Soil amendments:** Soil amendments like ground limestone which is used generally to correct soil acidity, also supplies available calcium. Similarly

gypsum and other sulphur containing compounds used to correct soil alkalinity also supply available calcium and sulphur to the plants.

5. **Some fungicides and weedicides:** Such as copper fungicides and Bordeaux mixture when used as fungicide supply available copper and calcium.

Exercise 1: A farmer of eastern Uttar Pradesh wants to apply 150 kg of nitrogen per hectare to the hybrid rice crop. How much amounts of the following fertilizers would be required?

- Ammonium sulphate (20% N)
- Ammonium chloride (25 % N)
- Urea (46 % N)
- Calcium ammonium nitrate (26% N)

Calculations:

$$\text{Amount of fertilizer required} = \frac{\text{Rate of fertilizer application (kg/ha)}}{\text{Content of nutrients in fertilizer (\%)}} \times 100$$

Amount of ammonium sulphate required = 150/20 × 100 = 750 kg/ha

Amount of ammonium chloride required = 150/25 × 100 = 600 kg/ha

Amount of urea required = 150/46 × 100 = 326.1 kg/ha

Amount of calcium ammonium nitrate required = 150/26 × 100 = 576.9 kg/ha

Exercise 2: A progressive farmer of western Uttar Pradesh is willing to grow 20 hectare wheat. Calculate the quantity of urea, single super phosphate and muriate of potash containing 46 % N, 16 % P_2O_5 and 60 % K_2O, respectively. The recommended dose of N, P_2O_5 and K_2O for wheat crop is 120, 60 and 40 kg/ha, respectively.

$$\text{Amount of fertilizer required} = \frac{\text{Rate of fertilizer application (kg/ha)}}{\text{Content of nutrients in fertilizer (\%)}} \times 100 \times \text{Area (ha)}$$

Amount of urea required = 120/46 × 100 × 20 = 5217.39 kg/ha

Amount of signle super phosphate required = 60/16 × 100 × 20 = 7500 kg/ha

Amount of muriate of potash required = 40/60 × 100 × 20 = 1333.33 kg/ha

Exercise 3: A farmer in his wheat crop wants to apply ntrogen in 3 equal splits as one-third at sowing, one-third at Crown root initiation and one-third at before panicle initiation stage. Recommended dose of N for dwarf wheat is 120 kg /ha. For basal dose calcium ammonium nitrate and for top dresing urea is available. Work out the quantity of these fertilizers for one hectare of wheat crop.

Total N requirement for dwarf wheat is 120 kg/ha and it has to be given in 3 equal splits. Therefore, at each application 40 kg N/ha is required.

Basal dose has to be given through CAN.

Amount of CAN required = 40/26 × 100 = 153.85 kg/ha

Amount of urea required = 40/46 × 100

= 86.96 kg/ha urea for top dressing each at CRI and before panicle initiation stage, respectively

Exercise 4: A farmer of Punjab has to apply 150 kg N/ha, 80 kg P_2O_5 and 60 kg K_2O/ha to a high yielding wheat variety of HD 2967. The available fertilizers are DAP, urea and muriate of potash. What quantity of each fertilizer required to meet out the requirement of wheat?

DAP contains = 18% N + 46 % P_2O_5, Urea = 46 % and MOP = 60 %

Amount of DAP required = 80/46 × 100 = 173.91 kg DAP/ha for P_2O_5

When we apply 173.91 DAP some amount of N is also supplied.

N supplied through DAP = 173.91/100 × 18 = 31.30 kg N

Requirement of N through urea (kg/ha) = Total N required – N supplied through DAP

= 150.00 – 31.30 = 118.70 kg

Amount of urea required = 118.70/46 × 100 = 258.04 kg/ha

Amount of MOP required = 60/60 × 100 = 100 kg MOP/ha

Exercise 5: A farmer wants to apply 80, 50 and 40 kg N, P_2O_5 and K_2O, respectively to his maize crop. He has well rotten Nadep compost with 0.80 % N, 0.5% P_2O_5 and 2.25% K_2O. For supplementating compost, urea and SSP can be purchased. Work out the dose of compost and other inorganic fertilizer for his 12.5 hectare land.

Note: Farmer is having well rotten Nadep compost and he can purchase urea and SSP i.e. no potassic fertilizer is available. It means that he has to apply entire potassium requirement from compost:

Total potassium requirement will be: 40 × 12.5 = 500 kg K_2O

Amount of compost requirement = 500 × 100/2.25 = 22222.22 kg compost for whole farm

= 22.22 tonnes

Amount of N supplied through compost: 22222 × 0.80/100 = 177.78 kg

Total amount of N required for 12.5 ha farm = 80 × 12.5 = 1000 kg N

Remaining amount of N = Total amount of N required – N supplied through compost

= 1000 – 177.78 = 822.22 kg N

Amount of urea required = 822.22 × 100/46 =1787.43 kg urea

P supplied through compost = 22222.22 × 0.5/100 = 111.11 kg P_2O_5

Total requirement of P_2O_5 = 50 × 12.5 = 625 kg P_2O_5

Remaining amount of P_2O_5= 625 – 111.11 = 513.89 kg P_2O_5

Amount of SSP required = 513.89 × 100/16 = 3211.8 kg

Exercise 6: Work out the cost per kg of nitrogen for ammonium sulphate and urea, if the prices for ammonium sulphate and urea are Rs. 3700 and 4600 per tonne, respectively.

Since ammonium sulphate contains = 20% N

Since the cost of 1000 kg (1 tonne) ammonium sulphate is = Rs. 3700

Therefore, cost of 1 kg ammonium sulphate will be = 3700/1000 = Rs 3.70/kg

Since, in 1 kg ammonium sulphate N is present = 0.2 kg

Since, the cost of 0.2 kg N is = Rs 3.7

Therefore, the cost of 1.0 kg N is = 3.7 × 1.0/0.2 = Rs.18.5/kg

Similarly, the cost of 1000 kg (1 tonne) urea is = Rs.4600

Therefore, the cost of 1.0 kg urea is = 4600/1000 = Rs.4.6/kg

Since, in 1 kg urea N is present = 0.46 kg

Since, the cost of 0.46 kg N is = Rs 4.6

Therefore, the cost of 1.0 kg N is = 4.6 × 1.0/0.46 = Rs.10.0/kg

The price of N supplied through urea per kg is cheaper than N supplied through ammonium sulphate.

14. SOIL FERTILITY AND PRODUCTIVITY

Soil fertility: Soil fertility is the ability of the soil to provide all essential plant nutrients in available forms and in a suitable balance.

or

Soil fertility denote the capacity of the soil to produce crops of economic value and maintain the health of the soil for future use.

or

It is the capacity of soil to supply essential nutrients to normal plants in adequate amounts and in a balanced proportion.

or

It is better to cultivate small piece of fertile land than large nutrient needs of the crop.

or

The soil is said to be fertile when it contains all sixteen of the required nutrients in the right proportion for luxuriant plant growth.

A productive soil in one in which the physical, chemical and biological conditions are favourable for plant growth.

Soil productivity: The capability of soil to produce specified crop yield under well defined and specified systems of management of inputs and environmental conditions.

Soil productivity factors

(A) Climate

1. Precipitation
2. Temperature
3. Humidity
4. Light

(B) Soil

1. Soil Organic matter
2. Soil Texture
3. Soil Structure
4. Soil Cation exchange

5. Altitude
6. Wind
7. CO_2

5. Soil Slope and Topography
6. Soil Temperature
7. Soil Water regine
8. Soil depth

Factors governing soil fertility

1. **Parent material:** Soil fertility depends on the chemical composition of parent material form which it derived.
2. **Topography:** Soils on the upper slope are less fertile than the soils on lower slope because of high leaching and erosion on upper slope.
3. **Climate:** In tropical climate decomposition of organic matter is faster than temperate climate. Thus soils of tropical regions are less fertile when compared to temperate region.
4. **Depth of soil profile:** Deep soils are more fertile than the shallow soils and the roots are spread well enough in deep soils than the shallow soils.
5. **Physical condition of soil:** The soil texture and structure influence the soil fertility.
6. **Artificial factors**
   (i) Waterlogging
   (ii) Cropping system
   (iii) Toxic chemicals and pesticides in the soil

Difference between soil fertility and productivity

Soil fertility	Soil productivity
1. It is an index of available nutrient to plants.	1. It is used to indicate crop yields.
2. Influenced by the physical, chemical and biological factors of the soil.	2. Depends upon fertility and location.
3. It is the function of available nutrients of the soil.	3. It is the function of soil fertility management and climate.
4. All fertile soils are not productive.	4. All productive soils are fertile.
5. It is an inherent properly of the soil.	5. It is not the inherent property of the soil.
6. Soil fertility can be estimated by yield of crops.	6. The quantity of biomass of plants produced each year on a given area (g/m^2)
7. Soil fertility can be analysed by chemical methods.	7. It can not be estimated by chemical methods.

Soil fertility evaluation

It is the assessment of nutrient supplying capacity of the soil.

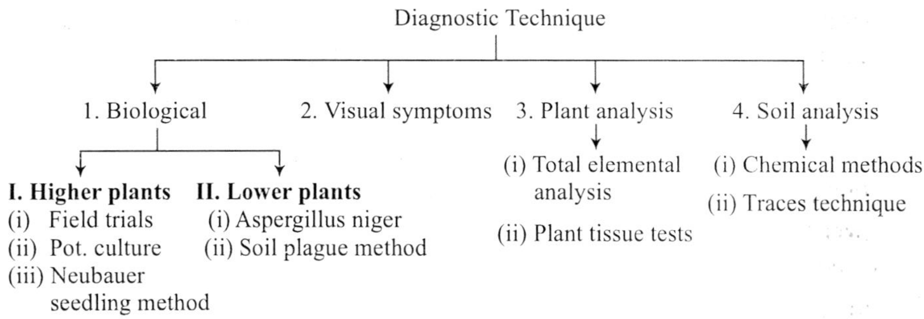

Biological methods: Using higher plants as indicator.

(I) Field trials: Direct way to assess the nutrient status in soil.

- Trials with graded doses of nutrients will determine the exact requirement.
- The effect is very much location specific.
- Expensive and time consuming and growth condition can not be fully controlled.

(II) Post culture method

- Same as field trials except that plants grown in pots with small volume of soil or under controlled field condition like green house.
- Limitations preferential root growth limited volume of soil problems during packing of soil.

(III) Neubauer seedling method: Soil is exhausted of the available nutrients and entire plant is then analysed for the uptake.

1. Biological methods: Using micro organisms as indicator.

(I) Asperigillus niger method

- For P, K and Mg in soils.
- The amount of K extracted in mycelium compares favourably with the content of exchangeable K in soil under investigation.

(II) Soil plaque method: By Sockett and stewart for study of mineral deficiencies in soil.

- It soil is deficient in K or P or both, the calories of the Azotobactor will not develop.
- The intensity of growth of calories indicates the degree of deficiency.

2. Visual symptoms

- Indicator plants
- Hidden hunger

3. Plant analysis method

(I) Total elemental analysis

- Leaf analysis widdy used approach
- Visual symptoms
- Deficiency

- Toxicity
- **Critical nutrient level (CNL):** CNL is a range of concentration at which growth of plants is restricted in comparison with that of plant at a higher nutrient level.

(II) Plant tissue tests Rapid chemical analysis

- The conductive tissue consisting un assimilate nutrients in sap or extracts are analysed.
- N, P and K have been tested by this technique.

4. Soil analysis

(I) Chemical method

- Phosphorus – Olsen's method
- Micro nutrient – Chlate DTPA
- Gypsum requirement – Schoonover's method

(II) Tracer technique

- The plants of few supplied the phosphatic fertiliser with P.
- The harvested plants listed for total P as well as 32P

$$A = B(1 - Y/Y)$$

where, A = Available P in soil

B = Amount of fertilizer P

Y = Fraction of P in plant desired from fertilizer P

(1 – Y) = Fraction desired from soil

15. TRANSFORMATION OF NUTRIENTS

The nutrients present in the soil are subjected to physical, chemical and biological change or transformations. During these transformations several substances are produced which are not directly connected to plant nutrition. Some of these substances may be toxic and others may be nontoxic. These transformations may either release or fix nutrients.

Nitrogen Transformation: During mineralisation, organic matter releases ammonium and nitrate which is available to plants. Nitrate may be further transformed into molecular nitrogen and escape into the atmosphere. The optimum soil moisture for mineralisation is occur at 50 to 70 per cent water holding capacity of soils. Mineralisation is temperature dependent process and rate of mineralisation is more at high temperatures and less at lower tepmerature.

Fertilizer materials added to the soils are transformed. After application of urea, it is hydrolysed in the presence of urease and forms ammonium carbonate. Most of the cultivated soils have enough urease to cause rapid hydrolysis.

$$\underset{\text{Urea}}{CO(NH_2)_2} + H_2O \xrightarrow{\text{Urease}} \underset{\text{Ammonium carbonate}}{(NH_4)_2CO_3}$$

Ammonium carbonate is an unstable compound and decomposes into ammonium and carbon dioxide. Ammonium is adsorbed on the clay complex, a portion is absorbed by the crop and around 11 per cent is lost as volatilisation. Since urea is soluble, leaching losses occur under lowland conditions. Therefore, mixing urea with soil is converted into ammoniacal ion and adsorbed on the complex. This reduces leaching losses. In alkaline soils, this method increases volatilisation. The applied and native forms of nitrogen are subjected to several losses. They are volatilisation, leaching, denitrification, ammonium fixation and immobilisation. Urea and ammoniacal fertilizers are subjected to volatilisation. Urea hydrolyses and ammonium carbonate is formed which dissociates and increases the pH around urea granules. The increase in pH initiates volatilisation losses.

$$NH_4 + H_2O \longrightarrow NH_3$$

Volatilisation of ammonia is maintained by soil alkalinity. In the case of ammonium fertilizers, external source of alkalinity *i.e.*, soil alkalinity, is necessary for initiating volatilisation. Ammonium ions react with water and hydroxyl ions and form ammonia and water. Ammonia, gas escapes into the atmosphere.

$$NH_4^+ + OH^- + H_2O \longrightarrow NH_3\uparrow + 2H_2O$$

Volatilisation losses range from 0.1 to 20.0 per cent and can be reduced by incorporation of fertilizers or by placement. Loss of nutrients beyond root zone along with water is known as leaching. Nitrate forms of nitrogenous fertilisers are subjected to higher leaching losses compared to other forms. The leaching losses with urea are around 7.5 per cent of applied nitrogen where urease activity is less. They are less with ammoniacal fertilisers as ammonium ions are adsorbed on clay complex. Leaching losses of these fertilisers occur in submerged soils and in soils with less CEC. The leaching losses that occur in a 12 week period from different fertiliser sources applied in 3 splits in sandy soils are high with ammonium nitrate (19.5 per cent) and lowest with urea (7.4 per cent). The leaching losses during this period are around 9.7 per cent with ammonium sulphate nitrate, 9.4 per cent with calcium ammonium nitrate and 7.7 per cent with ammonium sulphate.

Denitrificaton losses are more in submerged soils. Fixation of ammonium is due to trapping of these ions on montmorillonite, illite, and vermiculite minerals. They fix more ammonium when the soil is dry due to contraction of these minerals. On expansion, NH_4^+ is released by exchanging with Ca^{++}, Mg^{++}, Na^+ and H^+ ions. Fixed ammonium is available to microorganisms which convert them to nitrate and also to some plants which absorb them. The amount of nitrogen fixed ranges from 4 to 54 per cent of total N content.

Mineralisation is faster in aerobic soils which is carried out by fungi, actinomycetes and bacteria. When large quantities of organic matter is added to the soils, the microorganisms utilise available nitrogen in the soil. This temporary locking up of nitrogen in microorganisms is called immobilisation. However, it reverse process is known as mineralisation.

Phosphorus Transformation: It is present in the soils in organic and inorganic forms. Inorganic phosphorous is more than organic phosphorus. Phosphorus in soil solution depends on rate of decomposition of organic matter and rate of reaction with inorganic fraction. Mineralisation of organic matter release phosphorus in available form. The inorganic forms of soil phosphorus are Ca-P, Fe-P and Al-P. The dominant form of phosphorus in vertisols is Ca-P and in alfisols Fe-p. In Indian soils, Ca-P is around 40 to 50 per cent of total phosphorus in neutral and alkaline soils and more than 50 per cent in alkaline soils. Fe-P and Al-P are less than 10 per cent in these soils. Primary orthophosphate ($H_2PO_4^-$) is more in acid soils and secondary orthophosphate (HPO_4^{2-}) more in alkaline soils. Organic and inorganic phosphorus is in equilibrium with phosphorus in soil solution.

Solable P ⇌ Solution P ⇌ Insoluble Inorganic P

When fertilizers, containing soluble phosphates are added to the soil, phosphorus is retained or fixed by different mechanisms. Fixation of phosphorus is by adsorption, isomorphous substitution and double decomposition. Insoluble phosphates are weathered and phosphates are released into the soil solution. Thus equilibrium iis maintained. Phosphorus equilibrium is temporarily altered by addition of soluble fertilizers or by rapid mineralisation due to cultivation. Acidic soils fix more phosphates than neutral, alkaline and calcareous soils. It occurs in two ways: (1) by replacing two hydroxyl ions from aluminium in clay with one ion of primary orthophosphate, and (2) by forming clay phosphate linkage. At pH 2 to 5, fixation is mainly by formation of insoluble Fe and Al phosphates as Fe and Al are more active in acid soils. Within the pH range of 4.5 to 7.5, phosphate is fixed on clay particles.

When the clay content is high phosphorus fixation is more. Vermiculite and smectite clay fix more phosphates than kaolinite. Red soils have lesser fixing capacity than black soils. Aluminium oxides are more important than iron oxides in fixation. Phosphate fixation is high in soils with high calcium carbonate. Soluble phosphates are first converted into tricalcium phosphates and subsequently to carbonate apatite. The exchangeable calcium forms clay-calcium phosphate linkage or calcium enters soil solution and combines with phsphate. The solubility of P present in leachates of different soils is averaged to 1.4 mg/L. The solubility of P present in runoff water from high P soils is 0.11 mg/L. The presence of FYM shortens the period of P fixation and promotes its availability. In the presence of farmyard manure, P availability decreased during the first 6 weeks and then showed a considerable increase from the 8th week onwards. P fixation increased for up to 8 weeks with the rates of P in the absence of manure. With manure application, P fixation increased only during the first 6 weeks and thereafter decreased continuously. Alfisols has maximum P-fixation capacity (92.7%) followed by vertisols (86.5%) and inceptisols (76.6%) at 50 mg/kg of added P. Availabie P in soils generally increased over years due to cultivation of crops with phosphate fertilizers. Mineralization of P is more influenced by soil moisture than temperature. Application N fertilizer increases organic P in soil.

Potassium Transformation: Water soluble potassium is present in the soil solution as K^+ ion. It is in equilibrium with exchangeable K. Exchangeable K is present on soil clay complex which is adsorbed. Fixed potassium is bound between the units of clay minerals. These three forms *viz.* fixed, exchangeable and water soluble K are in dynamic equilibrium in the soil.

$$\boxed{\text{Fixed K.}} \rightleftharpoons \boxed{\text{Exchangeable K}} \rightleftharpoons \boxed{\text{Water soluble K}}$$

If one form is depleted, it is replenished from the other forms. Unavailable K or fixed K in mica and feldspars are made available slowly due to the weathering process;

$$KAlSi_3O_8 + H_2O \longrightarrow KOH + HAlSi_3O_3$$

Exchangeable K adsorbed on clay complex is released into soil solution when its concentration is decreased due to plant absorption. When fertiliser is added, K in the soil solution is increased. A fraction of it goes to exchangeable form and in due course even to fixed form.

Calcium Transformation: The important sources of calcium are dolamite, calcite, apatite and calcium feldspars. On their disintegration and decomposition, calcium is made available to plants. It may be taken up by plants, lost in drainage, re-adsorbed by clay or a small fraction may be reprecipitated as secondary calcium compound. Calcium is not fixed or made unavailable in the soil. Calcium in soil solution and in exchangeable complex are in a state of dynamic equilibrium. Higher concentration of calcium in solution is adsorbed on clay complex. Exchangeable calcium is released into soil solution when its concentration is low in soil solution.

Magnesium Transformation: Magnesium is available to plants due to weathering of biotite, dolamite, chlorate, serpentine and olivine. The released magnesium may be absorbed by plants and microorganisms, lost in drainage or reprecipitated as secondary minerals. Magnesium in soil solution and exchangeable form are in dynamic equilibrium, similar to calcium.

Sulphur Transformation: Sulphur is present as sulphides, sulphates and in organic combination with carbon and nitrogen. Earth's crust contains about 0.06 per cent of sulphur. The original source of sulphur in soil is sulphides of metals inplutonic rocks. During weathering process, these sulphides are oxidised to sulphates. These sulphates are present as insoluble sulphate salts in arid and semiarid regions.

The nature of applied sulphur is in the form of sulphates of Ca, Mg, K, Na or NH_4 in soil solution in arable soils. These sulphates may be adsorbed on 1 : 1 clay or hydrous oxides of Fe and Al. It may be absorbed by plants and microorganisms. Sulphates are reduced to sulphides in waterlogged soils and form H_2S, FeS etc. Elemental sulphur is oxidised to sulphates by microorganisms in aerated soils.

Transformation of Iron: It occurs in soil as oxides, hydroxides and phosphates in primary and secondary minerals. A small amount of iron is released due to weathering of primary and secondary minerals. Iron deficiency occurs in calcareous soils and soils with high phosphorus. The available and exchangeable iron decreases with increase in pH as it is inversely related to solubility of iron. Higher concentration of phosphorus causes deposition of iron on the root surfaces or just inside the root, possibly as iron phosphate. Iron is oxidised in the rhizosphere. Grasses have more oxidising power than legumes. Organic matter, native or added, improves the availability of iron. Iron availability is reduced with oxidation. Its availability is higher in submerged soils due to reduction of insoluble ferric compounds to soluble ferrous compounds.

Manganese Transformation: The primary minerals of manganese are present chiefly as oxides and to some extent as carbonates and silicates. The predominant secondary minerals are pyrolusite and manganite. Manganese in the soil is generally considered to exist in divalent (Mn^{++}), trivalent (Mn_2O_3) and tetravalent (MnO_2) forms in equilibrium.

$$\text{Exchangeable } Mn^{++} \rightleftharpoons MnO_2$$

$$Mn_2O_3$$

In acid soils, the available manganese is up to 16 per cent of total manganese content of soils, while in alkaline soils it seldom exceeds one per cent. Availability of manganese increases in acid soils as in the case of iron. Its availability is more in submerged soil due to reduction of insoluble manganic compounds to soluble manganous compounds. Tri- and tetra-valent forms are present under high pH and oxidising conditions. Organic matter decreases the availability of manganese. Calcium carbonate decreases manganese availability due to strong binding of Mn^{++} on calcium carbonate.

Transformation of Copper: Copper in earth's crust occurs mainly as sulphides and most abundant mineral of copper is chalcopyrite ($Cu\ Fe\ S_2$), It is also present in secondary minerals and organic complex. Weathering process releases copper. The solubility of applied Dpper fertiliser decreases with a rise in soil pH. Copper is fixed strongly on organic matter. It is due to adsorption of Cu^{++} on soil colloids and due to precipitation as copper hydroxides.

Transformation of Zinc: It is found chiefly as sphalerite, a sulphide minerals. It also exists as a constituent of organic matter. When soluble zinc salt is added to the soil, it undergoes several transformations by adsorption, precipitation, entry into clay lattice and by microbial consumption. The transformations are mainly influenced by soil pH. Within 24 hours after application of zinc fertilisers, only two per cent soluble zinc is present in soil solution in soils with pH above 7.0. The adsorption by clay complex and organic matter is high at higher pH. Generally, zinc availability is more in acid soils. Its availability is reduced with higher level of phosphates due to low solubility of zinc phosphate and also due to high organic matter as zinc is strongly adsorbed on humus. The per cent recovery of applied zinc increases on application of nitrogen. Similarly, application of potassium increases the water soluble plus exchangeable fractions of native soil Zn. Application of K increase different fractions of native soil Zn. Application of K also increases the transformation of applied Zn into all three fractions of the element in soil. K application caused an increase in Zn uptake and percentage utilization of both native and added Zn by rice. Potassium application increases the Zn uptake and per cent utilization of both native and applied Zn by lowland rice. Zinc deficiency is common in calcareous soils due to adsorption of zinc on carbonates of calcium and magnesium and formation of insoluble carbonates.

Transformation of Boron: As it is resistant to weathering, only a small fraction of boron is available to the plants from this source. Tourmaline, a fluorine bromosilicate is the important boron containing mineral. Most of the boron in cultivated fields is from organic matter. Mineralisation of organic matter releases boron, some of it is leached and some is held on clay. Irrigation water is another source of boron for crops. Boron added to the soil remains mostly soluble in acid soils and is subjected to leaching especially in light textured soils. Boron adsorption increases with pH. More than 72 per cent of applied water soluble boron is converted into water insoluble at pH above 7.5. The availability of applied boron is, therefore, reduced in alkaline soils.

Transformation of Molybdenum: It is present largely in the crystal lattice of primary and secondary minerals. Organic matter is another source of molybdenum. It is present in the soil as MoO_4^{2-}. It is either adsorbed on clay complex or present in the soil solution. Higher level of P fertilisers increases availability of molybdenum, but S reduces it. Unlike other micronutrients, its availability is high at high pH. MoO_4^{2-} is exchanged for two hydroxyl ions from the clay. Oxides of Fe and Al increase the adsorption of molybdenum. The ions of MoO_4 and SO_4 are similar in size and charge and they compete at root surface" for entry into the plant. Higher soil moisture increases molybdenum availability due to decrease in ferric iron which is responsible for adsorption.

Transformation of Chlorine: Chlorine is present in soils as a chloride and it is highly mobile in soil. It is not adsorbed on clay complex.

16. ASSIMILATION OF NUTRIENTS

Metabolic transformation of inorganic plant nutrients into organic plant constituents is known as assimilation. Assimilation literally means 'to make similar' i.e., the dissimilar carbon dioxide, water, nitrate, etc., are made similar to compounds found in plants. Carbon dioxide is metabolically transformed into carbohydrates by photosynthetic process. Water is assimilated in a few metabolic reactions, but bulk of it is lost in transpiration. Some nutrients undergo intensive metabolic transformations during their conversion from inorganic form whereas others do not. A fraction of the absorbed nutrients may be stored in vacuoles without being assimilated.

Assimilation of Nitrogen

Plants absorb nitrogen mostly as NO_3^- and NH_4^+. Most of the nitrate absorbed by roots translocated to leaves by transpiration stream. Nitrate assimilation is, therefore, mainly carried out in leaves and a small fraction in roots.

Nitrate io Ammonia: Nitrate in plants is first transformed into ammonia in a two step process. First, nitrate is reduced to nitrite (NO_2) and in this reaction, one atom of oxygen is released as water.

$$NO_3 + NADH + H^+ \xrightarrow{\text{Nitrate reductase}} NO_2^- + NAD^+ + H_2O$$

This process is catalysed by the enzyme nitrate reductase and the reducing power is supplied by NADH. As molybdenum is a constituent of nitrate reductase, deficiency of molybdenum blocks nitrate assimilation resulting in nitrate accumulation. Deficiency of molybdenum, therefore, causes nitrogen deficiency symptoms in addition to molybdenum deficiency symptoms. The second step is the reduction of nitrite to ammonia in the presence of nitrite reductase enzyme.

$$NO_2^- + 6e^- + 7H^+ \xrightarrow{\text{Nitrate reductase}} NH_3 + H_2O$$

In green leaves, the reaction occurs in chloroplasts and the reducing power is directly derived from light reactions. In other words, a portion of energy released during light reaction of photosynthesis is used for the reduction of nitrite to ammonia. Nitrite reduction occurs in the dark also both in root and leaf and the reducing power is supplied by respiration.

C4 plants assimilate nitrate more efficiently than C3 plants. Carbon dioxide reduction in C4 plants takes place in the bundle-sheath cells whereas light reactions and nitrite reduction takes places in mesophyll cells. The reducing power generated in mesophyll cells is not neccessary for carbon dioxide reduction in C4 plants. In C3 plants like rice, both CO_2 and NO_3^- reduction takes place in mesophyll cells. There is, therefore, competition for reducing power.

Ammonia to Amino Acids: The presence of ammonia in plants is due to reduction of nitrate to ammonia, absorption of ammoniacal nitrogen, breakdown of proteins during different physiological process, breakdown of urea absorbed by plants into ammonia and also by nitrogen fixation. Ammonia is assimilated rapidly in the plants. In most plants only traces of ammonia are present. Glutamic acid combine with ammonia to form glutamine. Glutamine in turn combines with ketoglutaric acid to form two molecules of glutamic acid which is an amino acid. Glutamic acid is used as base material for the synthesis of other amino acids and this process is known as transamination.

$$NH_3 + \text{Glutamic acid} + \text{ATP} \xrightarrow{\text{Glutamic synthetase}} \text{Glutamine} + \text{ADP}$$

$$\text{Glutamine} + \text{Alpha ketoglutaric acid} \xrightarrow[\text{synthetase}]{\text{Glutamic acid}} 2\ \text{Glutamic acid}$$

Most of the ammonia absorbed by plants is transformed into glutamine in root cells and transported to leaves where it combines with alpha ketoglutaric acid to form glutamic acid.

Assimilation of Phosphate

Phosphates absorbed by roots are partly translocated to leaves and a fraction remains the rocks. The fraction remaining in the roots is assimilated by oxidative phospholylation using the energy released during respiration.

$$\text{ADP} + \text{Pi} \xrightarrow{\text{Respiratory energy}} \text{ATP}$$

where Pi is inorganic phosphate.

The phosphate translocated to leaves is assimilated by photosynthetic phosphoiylation in chloroplasts by utilizing the energy released in light reaction.

$$ADP + Pi \xrightarrow[\text{Chloroplasts}]{\text{Radiant energy}} ATP$$

Phsophate assimilated in ATP is rapidly transferred by subsequent metabolic reactions into a wide variety of phosphorylated plant constituents viz. sugar phosphates, phospholipids, nucleotides etc.

Assimilation of Sulphate

Assimilation of sulphate takes place in chloroplasts using the energy released during light reactions

$$\text{Sulphate} \longrightarrow \text{Sulphite} \longrightarrow \text{Sulphide} \longrightarrow \text{Cysteine}$$

Sulphate ion reacts with ATP and is activated. It is reduced to sulphite and with the help of ferrodoxin, it is reduced to sulphide which in turn is incorporated into cysteine, a sulphur containing amino acid.

Assimilation of Cations

Cations *viz*., K, Ca, Mg, Fe, Mn, Cu, Na, etc., are assimilated either by weak inter ionic attractive forces of dissolved organic or inorganic anions or by formation of coordinated bonds. Most of the K and part of the Ca and Mg are attracted to anions in solution through their electric field force. The assimilation of metallic micronutrient cation Fe, Mn, Zn, Cu, Mo takes place through the formation of coordinated bonds. A large number of substances in plant cells are capable of forming coordinated bonds with metals including about one-quarter of all enzymes in plant cells. Some of them are phosphotases, chlorophyll, photosystem II, cytochromes, iron, sulphur, proteins etc.

17. MIXED FERTILIZER

Fertilizer materials containing two or more plant nutrient physically mixed manually or mechanically are called mixed fertilizers. Those fertilizer mixture containing all three major plant nutrients i.e. nitrogen, phosphorus and potash are known as complete mixed fertilizers, while those containing one or two plant nutrients are known as incomplete mixed fertilizers. In fertilizer mixture 2 or 3 straight (single) fertilizers are mixed thoroughly. For example, Harabahar (18 N : 18 P : 6 K) and sufla (20 N : 20 P : 0 K). The two types of mixed fertilizers (Farm made and Factory made) are prepared.

Principles of mixing fertilizers

The compatibility of various fertilizers for mixing depends upon the following reactions that take place.

1. **Loss of nitrogen from injudicious fertilizer mixture :** Basic compounds liberate ammonia gas from ammonium salts and organic nitrogenous manures, and lead to loss of nitrogen. Quick lime, slaked lime, potassium carbonate, basic slag and cynamide, which are basic in nature, should not therefore, be mixed with ammonical fertilizers and organic manures.

2. **Lumps formation in incompatible mixtures:** Superphosphate is likely to form lumps when mixed with other fertilizers and kept over for some time. It reacts with nitrates and liberates nitric acid and makes the mixture sticky and inconvenient for handling.
3. **Unavailability of phosphate in wrong mixtures:** Substances like slaked lime react with superphosphate on mixing and soluble monocalcic phosphate is reverted to dicalcic and tricalcic forms (Which are insoluble and unavailable to plants).

Types of fertilizer mixture

There are two types of fertilizer mixture:

1. Open chain formula fertilizer mixture
2. Closed chain formula fertilizer mixture

1. **Open chain formula fertilizer mixture:** The producers write the percentage of available nutrients from those mixtures, thus, it facilitates the farmers to select the right type of grade for his crop and soil.
2. **Closed chain formula fertilizer mixture:** These mixtures do not indicate their grade (per cent available nutrients) and type of fertilizers mixed together. Thus, the farmers cannot choose a correct mixtures for their use in production of crops.

Classification of mixed fertilizers

Mixed fertilizers are classified into four groups:

1. **Low grade mixed fertilizers:** If the sum of all three plant nutrients (NPK) is less than 14 in fertilizer mixture then they are known as low grade mixed fertilizer. For example 3 – 4 – 3.
2. **Medium grade mixed fertilizers:** If the sum of all three plant nutrients (NPK) is 15 to 25 in fertilizer mixture then they are known as medium grade mixed fertilizer. For example 4 – 6 – 6.
3. **High grade mixed fertilizers:** If the sum of all three plant nutrients (NPK) is 25 to 35 in the fertilizer mixture then they are known as high grade mixed fertilizer. For example 11 – 11 – 11, 8 – 10 – 12.
4. **Concentrated grade mixed fertilizers:** If the sum of all three plant nutrients (NPK) is greater than 35 in the fertilizer mixture then they are known as concentrated grade mixed fertilizer.

Materials used in fertilizer mixtures:

The following types of materials are used in making the mixture:

1. **Nutrients:** Supplier of plant nutrients (fertilizer).
2. **Conditioner:** They help in maintaining a good physical condition of the mixture. They can be powder of tobacco stalks, paddy husk powder etc.

3. **Neutralizer:** They are used to neutralize the residual acidity of certain fertilizers in mixture e.g. ammonium sulphate. Mostly dolomite, limestone are used for this purpose.
4. **Filler:** They are sand, powdered coal, ash etc.

Advantages of fertilizer mixture

The advantages of fertilizer mixtures are summarized as follows:

1. The farmer is supplied a balanced fertilizer mixture that is adopted for each soil and each crop.
2. Time and labour costs are saved since all of the fertilizer materials are applied at one time.
3. Fertilizer mixtures can be prepared in a manner to correct residual acidity of certain nitrogen fertilizers.
4. Fertilizer mixtures in balanced form help in maintaining good physical condition of soil.
5. One nutrient increases the availability of other nutrient.
6. Balanced fertilizer mixtures provide tolerance against drought, cold, pests and diseases.
7. Fertilizer mixture can be stored easily.
8. The loss of nutrients through mineralization is minimum in mixed fertilizers.

Disadvantages of fertilizer mixture

The disadvantages of fertilizer mixture are as follows:

1. The unit cost of plant nutrients, in mixtures is higher than that of purchasing the same materials separately.
2. The improper mixtures may not be suitable for all soils.
3. The purchase and use of correct formulation of fertilizer mixture requires relatively greater knowledge. Lack of proper knowledge about the proper mixture for a specific soil type and for a particular crop may result in improper use.
4. When mixed fertilizers are used, single nutrient fertilizer may still be necessary at a specific plant growth stage. For example, nitrogen in split application (top dressing), i.e., at 2 or 3 stages of plant growth. Therefore, farmers have to purchase nitrogenous fertilizers for split application (top dressing), apart from mixed fertilizers.

Fertilizer mixtures available in the market

1. Sufla (15 : 15 : 15)
2. Sufla (20 ; 20 : 20)
3. Lakshmi (12 : 12 : 12)
4. Lakshmi (8 : 8 : 8)
5. IFFCCO-1 (10 : 26 : 26)
6. IFFCCO – 2 (12 : 32 : 16)

Grade:

Grade of mixed fertilizer is 5 – 5 – 10 (fertilizer ratio of 1 : 1 : 2), means that 100 kg of the mixed fertilizer of this grade will contain 5 kg of N, 5 kg of P_2O_5 and 10 kg of K_2O. The source of N, P and K are Urea (46% N), single super phosphate (16 % P_2O_5) and muriate of potash (60% K_2O), respectively. For this grade, 10.9 kg urea, 31.3 kg super phosphate and 16.7 kg muriate of potash will be used. The total weight of fertilizer ingredient will be 58.9 kg. To this fertilizer mixture, 41.1 kg of inert material will have to be added to make up 100 kg bag of mixed fertilizer.

18. ORGANIC FERTILIZERS

The organic manures are made up of dead plant and animal remains. They are bulky in nature and poor in plant food. They supply almost all the ingredients of plant food.

There are two types of organic manures:

1. Bulky organic manures are farm yard manures, compost, manures of night soil and green manures.
2. Concentrated organic manures are oilseed cakes, blood meal, bone meal etc. are rich in nutrients.

Composition of organic manures

Organic manures	Nutrient content (kg/ha)		
	N	P_2O_5	K_2O
Bulky organic manures			
Farmyard manure	0.50	0.25	0.50
Compost	0.89	0.51	3.05
Manure of night soil	1.00	0.80	0.40
Concentrated organic manure			
Groundnut cake	7.29	1.53	1.33
Linseed cake	5.56	1.44	1.28
Castor cake	4.37	1.85	1.39
Neem cake	5.22	1.08	1.48

Farmyard manures (FYM)

There are two methods used for preparation of farmyard manure as discussed below:

Heap method: Heap method of storing is recommended. The heaps are prepared on the upland. Each day' s material is arranged, layer by layer in dome shape. When the heap is about 6 ft high, it is round on the top and plastered with mud. After six months good quality of manure is prepared.

Trench method: The manure preparation should be carried out in trench of 20-35 ft length, 5-6 ft breadth and 3-31/2 ft deep. Dry litter should be spread under each animal in the evening for absorption of urine. Each morning urine-soaked litter and dung should be well-mixed and filled in the one section of 3 ft length of the trench. When one section is filled to a height of 11/2 – 2 ft above ground level, the dome shaped top surface is plastered with a cowdung and soil slurry. After filling the one section (3 ft) of trench, the next 3 ft length of the trench is filled. In about 3 months the manure is ready for use.

Compost

The process of decomposing organic wastes is called composting and the decomposed materials is called compost.

Bangalore or Acharya method : The composting of the vegetative waste is done in pits. The size of pit (9 × 2.1 × 0.9 metre) varies with the quantity of material available for composting. A 22 cm thick layer of waste material is first laid in pits. One each layer spread a slurry made of dung, urine earth. A sufficient quantity of water is sprinkled over the material in the pit to make it moist but not too wet. The pit is filled in this way layer by layer, the layer of bedding material alternating with that of slurry and water, till it rises 2 ft above ground level. At the end dome-shaped heap is plastered over with wet mud. It is left undisturbed. The compost is ready in about 4 to 5 months. This method overcomes many of disadvantages of the Indore method like turning etc. After the initial aerobic decomposition during the first eight to ten days (i.e., filling of pit is going on) the material undergoes semi-anaerobic decomposition. Under this process, there is less loss of organic matter and nitrogen and hence there is a greater production of compost.

NADEP compost: This method is developed by Nadep kaka of Maharastra. The method, which has become quite popular among the farmers in Western India. The Nadep method of making compost is unique not because it is successful in making good compost, which other methods can also be claim to, its real secret lies in the large quantities of compost the process can deliver with a minimum human effort within a specific period of time. The process basically involves placing select layers of different types of compostible materials in a simple, mud sealed structure designed with brick and mud water. The system permits conversion of approximately 1 kg of animal dung into 40 kg of rich compost which can be applied directly to the field.

Role of organic manures

Role of organic manures is given below:

1. Organic manure binds soil particles into structural units called aggregates. These aggregates help to maintain a loose, open, granular condition. Water infiltrates and percolates more rapidly. The granular condition of soil maintains favourable condition of aeration and permeability.
2. Water holding capacity is increased by organic matter. Organic matter definitely increases the amount of available water in sandy and loamy soils. Further, the granular soil resulting from organic matter additions, supplies more water than sticky and impervious soil.

3. Surface runoff and erosion are reduced by organic matter.
4. Organic matter or organic manure on the soil surface reduces losses of soil by wind erosion.
5. Surface mulching with coarse organic matter lowers soil temperatures in the summer and keeps soil warmer in winter.
6. The organic matter serves as a source of energy for the growth of soil microorganisms.
7. Organic matter serves as a reservoir of chemical elements that are essential for plant growth. Most of the soil nitrogen occurs in organic combination. Also a considerable quantity of phosphorus and sulphur exist in organic forms. Organic matter supplies the nutrients needed by growing plants, as well as many hormones and antibiotics.
8. Fresh organic matter has a special function in making soil phosphorus more readily available in acid soils.
9. Organic acids released from decomposing organic matter help to reduce alkalinity in soils.
10. Fresh organic matter supplies food for earthworms, ants and rodents. These microorganism improves drainage and aeration. Earthworms can flourish only in soils that are rich in organic matter.
11. Organic matter upon decomposition produces organic acids and corbondioxide which help to dissolve minerals such as potassium and make them more available to growing plants.
12. Humus (Highly decomposed organic matter) provides a storehouse for the exchangeable and available cations – potassium, calcium and magnesium. Ammonium fertilizers are also prevented from leaching because humus holds ammonium in an exchangeable and available form.
13. It acts as a buffering agent. Buffering checks rapid chemical changes in pH.

19. GREEN MANURING

Green manuring can be defined as a practice of ploughing or turning into the soil undecomposed green plant tissues for the purpose of improving physical condition as well as fertility of the soil.

Green manuring in situ: In this system, green manure crops are grown and buried in the same field which is to be green manured, either as a pure crop or as intercrop with the main crop. The important green manure crops are sunhemp (*Crotolaria juncea*), dhaincha (*Sesbania aculeata*), pillipesera (*Phasolus tribulus*), guar (*Cyamopsis tetragonoloba*), green gram (*Phaseolus vulgare*), *cowpea*, senji, berseem, *black gram* etc.

Leaf green manuring: Leaf green manuring reflects to turning in to the soil of green leaves and tender green twigs collected from shrubs and trees grown on bunds, wasteland and nearby forest areas. The common shrubs and trees used are Glyricidia

(*Glyricidia maculate)*, Subabool *(Sesbania speciosa)*, Karaj (*Pongamia pinnata*), neem (*Azadiracta indica*), etc.

Principles of green manuring

1. Green manure crop should be grown in irrigated area or where annual rainfall is more than 750 mm. Lack of moisture is harmful for the growth of the crop as well as for decomposition. An undecomposed crop may harm the subsequent crop by upsetting the balance of carbon and nitrogen.
2. It should be preferably from the legume family.
3. Green manure crop should be quick growing.
4. Green manure crop should have more foliage.
5. After green manuring subsequent crop should be sown in well decomposed crops. Undecomposed green manure may cause poor germination and problem of diseases and pests.
6. It can be grown in all types of soil.
7. It should have deep root system.
8. In irrigated area, the best stage at which the crop should be incorporated in the soil as a green manure is when it reaches the flowering stage. In rainfed or dry region, green manure crop should be incorporated before flowering stage.

Advantages of green manuring: There are following advantages of green manuring:

1. It increases the organic matter regime of the soil and thereby modifies soil physical, chemical and biological environments. In facts, this stimulates the activity of soil micro-organisms.
2. It reduces the soil loss caused by runoff and erosion.
3. It improves the soil structure, aeration status, permeability and infiltration capacity of soil.
4. The green manure crops help for returning the different plant nutrients from the surface soil layer to the sub-surface soil layer.
5. Green manure crops have some residual effect in relation to supply of different plant nutrient and thereby it helps for the better growth to the next crop.
6. Green manuring checks weed growth. The plants used for green manuring are usually, grow very quickly and thus, tend to suppress the growth of weeds.
7. Green manuring crops aid in the reclamation of saline and alkaline soils by the release of organic acids.

Disadvantages of green manuring

The limitations of green manuring are as follows:

1. Under rainfed conditions where rainfall is limiting, the proper decomposition of the green manure crops may not take place and thereby benefits of green manuring may not be achieved satisfactorily.

2. Sometimes the cost of green manuring is more than chemical fertilizers.
3. Due to improper decomposition, problems of insects-pests and diseases may come up.
4. If rainfall is scanty, growth of green manure crops would be less vigorous which results in less production of green matter.
5. Green manure crop may be taken as a catch crop between the main crops. Due to late sowing of green manure crop and insufficient moisture, burying of green manure crop becomes late. Therefore, sowing of main crop may not be done or delayed.
6. Due to decomposition of green manure crops, various toxic substances like organic acids e.g. butyric acid, propionic acid etc. and toxic gases like methane (CH_4) and others etc. are liberated which affect the root growth of growing plants and thereby affect the growth and yield through inhibition of nutrient absorption by the plant.

Selection of green manure crops

The characteristics of good green manuring crops are given below:

1. It should be quick growing, so that timely incorporation of green manure crops may be done. For example – sunhemp and green gram.
2. It should yield large quantities of green material in a short period. For example – dhaincha (*Sesbania aculeata*).
3. It should be preferably from the legume family so that nitrogen would be fixed in addition to green matter production. For example – dhaincha and sunhemp.
4. It should be tender (more leafy growth than woody growth) so that its decomposition will be rapid. For example – berseem and *lobia* (cowpea).
5. It should have a deep root system so that it would penetrates deep layers of the soil. Thus, it utilizes nutrients and water from deeper layers and also helps in developing good soil structure. For example – sunhemp and dhaincha.
6. Green manure crops have capacity to grow under limited moisture supply conditions.
7. Green manure crops should not be affected by any insects and pests and diseases.
8. The green manure crop should have capacity to decompose easily and faster.

Green manure crops their yield and nitrogen fixation

Name of green manure crops	Botanical name	Average green matter yield (q/ha)	Nitrogen fixation (kg/ha)
Sunhemp	*Crotolaria juncea*	194.7	85.4
Dhaincha	*Sesbania aculeata*	185.0	80.8
Black gram	*Phaseolus mungo*	95.5	45.5
Green gram	*Phaseolus aureus*	92.7	43.6
Khesari	*Lathyrus sativa*	125.0	61.5

Name of green manure crops	Botanical name	Average green matter yield (q/ha)	Nitrogen fixation (kg/ha)
Berseem	*Trifolium alenxandrinum*	170.5	58.7
Cowpea	*Vigna unguiculate*	220.0	95.6
Guar	*Cyamopsis tetragonoloba*	215.5	99.5

20. METHODS AND TIME OF FERTILIZER APPLICATION

Principles involved in selection of method of fertilizer application:

1. Nitrogen fertilizer is easily soluble in water and moves rapidly in all directions from place of application. Nitrogen applied in soil surface reaches the plant root easily and rapidly. As such, these fertilizers are broadcast on the soil surface just before sowing.
2. Phosphatic fertilizers are not much suitable for top dressing.
3. Leaching and volatilization loss is very high in nitrogenous fertilizers. So fertilizers are suitable for top dressing and side dressing.
4. To reduce the fixation of phosphate, phosphatic fertilizers should be placed to the plant roots. In other words, localized placement of phosphatic fertilizers near the seed or seedling roots should be practiced.

Methods of fertilizer application

Various methods are used for fertilizer application. The methods are classified as follows:

A. Application of fertilizers in solid form

1. Broadcast

(a) **Broadcasting:** Evenly spreading the dry fertilizers over the entire field and incorporation by cultivation is termed as broadcasting.

(b) **Basal application:** Evenly spreading of solid fertilizers over the entire field before or at sowing or planting is called basal dressing.

(c) **Top dressing:** The broadcasting of the fertilizer on closely sown standing crops is called top dressing. Generally nitrogenous fertilizers are used for top dressing. N – fertilizers should not be applied on wet leaves.

2. Placement

(a) **Plough sole placement:** Fertilizers is placed in a continuous band on the bottom of the furrow during the process of ploughing.

(b) **Deep placement:** Usually nitrogenous fertilizer (urea) is placed in reduction zone (deep in the soil), where it remains in ammonical form and is available to the crop slowly.

(c) **Subsoil placement:** Fertilizers like phosphate and potash are placed in sub soil. This method is recommended in humid and semi humid regions where many sub soils are strongly acidic.

3. Localized placement

(a) **Contact placement, combined –drilling or drill placement:** fertilizers are drilled alongwith the seed, using suitable equipment (e.g., fertilizer drills or seed – cum – fertilizer drill).

(b) **Band placement:** This is a system of localized placement of fertilizer in the hill or along the row of crops.

(c) **Pellet placement:** This is the methods of fertilizer application in the form of pellets particularly in the rice field.

(d) **Side-dressing:** The fertilizer applied after the crop has been seeded in spaced rows. This may be done either as a continuous band near the crop row or dibbled in between the plants. Side dressing is a very common practice in long duration crops such as sugarcane.

(e) **Spot placement:** Some time the fertilizer is placed in between the plants in pinches at the base of each pair of plants.

(f) **Ring placement:** In this method, fertilizers are applied around the plants in ring (basin).

B. Application of fertilizers in liquid form

(i) **Starter solution** – 1:2:1 or 1:1:2 ratio of NPK to vegetable crops at the time of transplanting; always use small quantity of fertilizer solution.

(ii) **Foliar application/spray:** Some fertilizers in liquid form can be applied by spraying directly on the leaves of crop plants. Several nutrient elements are readily absorbed by leaves. It can be sprayed at the rate of 2-4 per cent. Three kg of urea in 500 L of water has been reported safe. Urea and micronutrients like zinc sulphate and magnesium sulphate are common.

(iii) Direct application to soil

(iv) Application through irrigation water or ferti-irrigation.

Time of application of fertilizer

Principles governing selection of proper time for application of fertilizers are:

1. Nitrogen is required throughout the crop growth. Nitrogen is taken up by the plant slowly in the beginning, rapidly during the grand growth period and again slowly near maturity.
2. Nitrogen is lost easily through leaching. Therefore, it is better not to apply too much nitrogen fertilizer at one time, but to apply in split doses throughout the growth period.
3. Phosphorus is required during the early root development and early plant growth. As such, crop plants utilize $2/3^{rd}$ of the total requirement of phosphorus when the plants accumulate $1/3^{rd}$ of their dry weight.
4. All phosphatic fertilizers released phosphorus for plant growth slowly. As such, it is always recommended that the entire quantity of phosphatic fertilizers should be applied before sowing or planting.

5. Potash behaves partly like nitrogen and partly like phosphorus. From the point of view of the rate of absorption, it is like nitrogen, being absorbed, upto the harvesting stage. But potash fertilizer like phosphate becomes available slowly. As such, it is always advisable to apply the entire quantity of potash at sowing time.

21. ENHANCING FERTILIZER USE EFFICIENCY

There are several factors which influences the fertilizer use efficiency in crop production. These include:

(i) Type of soil and its fertility status.
(ii) Cropping history and previous crop grown
(iii) Season in which the crop is grown
(iv) Nature of crop, variety and its growth habit
(v) Sowing time and plant density
(vi) Type, quantity, time and method of fertilizer application
(vii) Method, quantity and frequency of irrigation
(viii) Consevation of moisture especially under dryland.

Following factors which enhances fertilizer use efficiency:

(i) By and large irrigated crops respond better to fertilizer as compared to non-irrigated crops.
(ii) Most responsive and best suited crops and their varieties for the locality should be selected to get maximum benefit of the limited quantity of fertilizers.
(iii) High yielding crop varieties always gave higher yield response to fertilizers than the local varieties at lower rates of application.
(iv) Under delayed sown conditions, the yield as well as the response per kg of nutrient used is reduced.
(v) Proper plant spacing has to ensure to get higher yield.
(vi) It is desirable to properly conserve and use of organic waste materials in the farm to build soil fertility and its water holding capacity as well as supplying plant nutrients to the crop for the longer period.
(vii) While using compost made from straw and leaves, it will be necessary to add a small quantity of nitrogen to the compost so that availability of nitrogen to the young crop is not adversely affected.
(viii) Introduction of leguminous crops in diverse rotational and inter-cropping sequence and use of bacteria and algal cultures play an important role in meeting the fertilizer need of the crop.
(ix) Leguminous crops have the unique capacity of not only fixing atmospheric nitrogen in the soil but also have the ability to mobilize non-available native soil phosphorus to the available form and improved soil physical, chemical and biological conditions.

(x) Before sowing seed should be treated with rhizobium culture and fungicides.

(xi) Always use healthy and disease free seed/seedling.

(xii) Green manuring with sunhemp and dhaincha should be practised in low land rice cultivation.

(xiii) Excess irrigation water should not be used. Always apply irrigation water at critical stages of crop growth.

(xiv) Balanced fertilization should be practised based on the soil test.

(xv) Fertilizer recommendation should preferably be based on crop sequence for multiple cropping.

(xvi) Nitrogen fertilizer should be applied in split doses, while phosphatic and potassic fertilizers as basal dressing.

(xvii) Under limited moisture conditions, 2% foliar spray of urea gives better results. For rice crop, if possible urea mixed with nitrification inhibitors such as neem coated etc.

(xviii) Apply micro nutrients such as zinc, boron, sulphur on the basis of soil test.

(xix) In acid or saline-alkali soil, soil amendments such as lime, gypsum etc. should be applied before using fertilizers and properly mixed in the soil.

(xx) Weeds deplete maximum quantity of plant nutrients than crop. It should be ensured that proper and timely weed control measures are adopted to create weed free environment to the crop.

(xxi) Proper control of insects/pests and disease is a must for realizing maximum effectiveness from fertilizers.

22. CLASSIFICATION OF FERTILIZERS

Classification of nitrogenous fertilizers: Nitrogen fertilizers are classified into 4 groups on the basis of chemical properties:

1. **Nitrate fertilizers:** Nitrogen is available in the form of nitrate or nitrite. These are

Sodium nitrate or chelean nitrate	16%N
Calcium nitrate	15.5 %N

2. **Ammonium fertilizers :** Nitrogen is available in the form of ammonia. These are:

Ammonium sulphate	20% N + 24 % S
Ammonium phosphate	20%N + 20% P_2O_5
	16% N + 20 % P_2O_5
Ammonium chloride	24-26 % N
Anhydrous ammonia	82 % N
Ammonia liquid	20-25 % N

3. **Nitrate and ammonium fertilizers :** Nitrogen is available in the form of nitrate and ammonia. These are:

Ammonium nitrate	33-34 %N

Calcium ammonium nitrate 26 % N

Ammonium sulphate nitrate 26 % N

4. **Amide fertilizers:** Nitrogen is available in the form of amide. These are:

Urea 46 %N

Calcium cyanamide 21 % N

Classification of phosphatic fertilizers

Phosphatic fertilizers can be classified into three groups depending on the form in which orthophosphoric acid or phosphoric acid is combined with calcium. These are:

1. **Water soluble phosphatic fertilizers or monocalcium phosphate:**

Single super phosphate 16-18 % P_2O_5

Double super phosphate 32 % P_2O_5

Triple super phosphate 46-48 % P_2O_5

Ammonium phosphate 20%N + 20 % P_2O_5

16%N + 20 % P_2O_5

2. **Citric acid soluble phosphatic fertilizers or dicalcium phosphate:**

Basic slage 14-18 % P_2O_5

Dicalcium phosphate 34-39 % P_2O_5

Rhenania phosphate 23-26 % P_2O_5

Raw and steamed bone meal

Part of P_2O_5 soluble in citric acid

3. **Water and citric acid insoluble phosphatic fertilizers or tricalcium phosphate:**

Rock phosphate 20-40 % P_2O_5

Raw bone meal 20-25 % P_2O_5 and 3-4 % N

Steamed bone meal 22 % P_2O_5

Classification of potassic fertilizers: Potassic fertilizers can be classified into:

1. **Fertilizers having K in the chloride form :** The most common fertilizers are:

Muriate of potash 60 % K_2O

2. **Fertilizers having K in non- chloride form**: The most common fertilizers are:

Sulphate of potash 48-52 % K_2O

Potassium nitrate 44 % K_2O + 13 % N

Potassium magnesium nitrate 20-30 % K_2O + 12 % N

(Potash magnesium)

23. FUNCTIONS OF ESSENTIALS PLANT NUTRIENTS

The functions of various plant nutrients are described below:

Nitrogen:

1. Nitrogen is an essential constituent of proteins and chlorophyll and is present in many other compounds of great physiological importance in plant metabolism such as nucleotides, phosphatides, alkaloids, enzymes, hormones, vitamins etc. Thus it is basic constituents of life.
2. Produces rapid early growth and imparts dark green colour to plants.
3. Increases protein content in food and fodder crops. Improves quality, succulence of leafy vegetables and fodder crops.
4. Promotes vegetative growth of plants.
5. Regulates utilization of potassium, phosphorus and other elements.

Phosphorus:

1. Phosphorus is a constituent of nucleic acid, phytin and phospho-lipids.
2. Phosphorus is also an essential constituents of majority of enzymes which are of great importance in the transformation of energy, carbohydrate metabolism and also in respiration of plant.
3. It is closely related to cell division and development.
4. Stimulates early root development and growth.
5. Rapid and vigorous plants growth, strengthens straw and decreases lodging tendency.
6. Brings early maturity of crops particularly in cereals.
7. Stimulates flowering and seed setting and increases harvest index.
8. Improves quality of food grains and other crops.
9. In legumes, it enhances the activity of rhizobia and increases the formation of root nodules thus nitrogen fixation enhances in many folds.

Potassium:

1. Imparts increased vigour and disease resistance to plants.
2. Produces strong and stiffness of straw in cereals (rice and wheat). Thus it reduces lodging.
3. Imparts winter hardiness to legumes and other crops.
4. Regulates water balance within the plant cell. Thus it reduces tendency to wilt and helps in better utilization of available water.
5. Helps in formation of protein and chlorophyll and accelerated enzyme action.
6. Plays essential in the formation and transfer of starch and sugar.
7. Increases plumpness of grains and seeds.
8. Improves the quality of tobacco leaf, fibre crops and taste, size and keeping quality of fruits.

Calcium:

1. It is constituents of the cell wall, as such, it increases stiffness of straw.
2. Essential to activate growing points especially root tips.
3. Influences the water economy of the plant, the protein-carbohydrate ratio in fat metabolism as well as many other physiological processes.
4. Improves intake of other plant nutrients such as iron, boron, zinc, copper and manganese.
5. Provides a base for neutralization of organic acids.
6. Promotes early root development and growth and improved quality and quantity of seed.

Magnesium:

1. It is a constituent of chlorophyll and it helps in maintaining the dark green colour in leaves.
2. Helps in translocation of starch and regulates the uptake of other nutrients.
3. Acts as a carrier of phosphorus in plants particularly in connection with the formation of seed of high oil content. Thus it promotes the formation of oils and fats.
4. Plays a part in the production of carbohydrates, proteins, fats and vitamins and in certain catalytic reactions in the enzyme systems.

Sulphur:

1. Though sulphur is not a constituent of chlorophyll, but it helps in chlorophyll formation and encourages vegetative plant growth.
2. Increases root growth and promotes root nodules in legumes.
3. It is an essential constituent of many proteins, enzymes and certain volatile compounds such as mustard oil and helps in oxidation-reduction in respiration and in the activation of ferments.
4. Stimulates seed formation.

Boron:

1. It is a constituent of cell membrane and essential for cell division.
2. It helps in translocation of sugar in plants and regulate potassium/calcium ratio in the plant.
3. The primary role of boron appears to be concerned with calcium metabolism, both with its uptake by roots and its efficient use in plants.
4. Helps in the absorption of nitrogen.
5. It affects many other functions in plant such as active salt absorption, hormone movement, flowering and fruiting processes, pollen germination, carbohydrate

as well as nitrogen metabolism, respiration, water metabolism and water relations in the plant.

6. Tends to keep calcium soluble and increases its mobility in the plant.

Manganese:

1. The function of manganese is closely associated with that of iron. It also supports the movement of iron in the plant.
2. Helps in the chlorophyll formation.
3. Acts as catalyst in oxidation and reduction reactions within the plant tissues.
4. It helps in counteracting the bad effect of poor aeration.

Iron:

1. Though it is not a constituent of chlorophyll, but it helps in its formation.
2. Helps in absorption of other plant nutrients.
3. It is essential for synthesis of protein content in the chloroplasts.
4. It regulates respiration, photosynthesis, reduction in nitrates and sulphates.

Zinc:

1. Zinc is a constituent of several enzymes which regulate various metabolic reactions in the plants i.e. oxidation-reduction reactions in the formation of chlorophyll.
2. Influences the formation of some growth hormones in the plant.
3. Helpful in reproduction of certain plants.
4. It is associated with water uptake and water relations in the plants.

Molybdenum:

1. Acts as enzyme systems which bring about oxidation-reduction reactions, especially the reduction of nitrates to ammonia prior to amino acid and protein synthesis in the cells of the plant.
2. It is an essential for the process of atmospheric nitrogen fixation, both symbiotic and asymbiotic.

Copper:

1. Copper forms many compounds with amino acids and proteins in the plant.
2. It acts as electron carrier in enzyme which brings about oxidation-reduction reactions in plants.
3. Helps in the utilization of iron in chlorophyll synthesis.

Chlorine:

It is needed for proper plant development especially in sugarbeet, carrots, lettuce, barley, wheat, cotton and clovers.

Elements	Major role
C, H, O	Provide basic structure
C, H, O, N, P, S	Tissue building elements
K, Mg, Ca, Cl	Electrolytic balance
Ca, Mg, P	Skeletal elements
H, O	Energy exchange elements
N, P, S	Energy storage, transfer and bonding
C, N, S, P	Plastic and storage of energy
Fe, Co, Mn, Cu, Zn	Oxidation-reduction (redox) regulators
Cu, Fe, Mn, S	Catalytic elements

24. DEFICIENCY SYMPTOMS OF PLANT NUTRIENT

- **Nitrogen:** Nitrogen is readily translocated from old to young leaves under stress conditions. The deficiency symptoms always appear first on the old leaves and the young leaves remain green for some time. The light green or pale yellow colour extending from the tip towards the base of the leaf including the mid rib and whole plant looks like pale. Plant shows poor tillering and shortened internodal length leading to poor growth and crop yield.
- **Phosphorus:** The deficiency appears first on old leaves. Phosphorus deficient plants show slow growth concomitant with dark green colour of leaves. Later the leaves turn purple from tip towards base and start drying. In acute deficiency, stress of the plants also turn into purple. Plants having phosphorus hunger show poor tillering and have relatively less height. Maturity is also delayed in phosphorus deficient plants leading to poor grain yield.
- **Potassium:** The deficiency of potassium is also first exhibited by old leaves of plants. The affected leaves appear faded out and develop chlorosis followed by the appearance of necrotic areas at the tip and along the margins of leaves extending inwards with time. Later because of scorching of these areas, they look brown and become brittle. Short supply of potassium to the plant may bear shriveled seeds. It reduces the resistance against insects/pests and diseases.
- **Sulphur:** Sulphur deficiency in crops first shows up on the young leaves. It is manifested as general yellowing of these leaves from a point slightly below the tip towards the base. With time, the entire portion of the leaf blade becomes yellow. Under severe deficiency, the whole plant appears chlorotic or pale yellow. The growth is restricted and the whole plant looks stunted.
- **Zinc:** In rice, symptoms appear as light yellow spots scattered in the interveinal area. These spots later turn yellowish – brown imparting pale yellowish brown colour to the affected leaves. The zinc deficient plants generally have shorter internodal length, less number of productive tillers, poor foliage expansion,

bushy growth, small ears and poor grain formation. Heading and maturity is also delayed.

- **Iron:** Deficiency of iron, is usually exhibited as interveinal chlorosis of the young or new leaves. Soon after the veins also loss of green colour and the whole leaf turns yellow. Under acute deficiency conditions, there is bleaching of the affected leaves and the newly emerging leaves also look white or bleached. In some crops reddish spots/lesions may also develop. With time the affected leaves turn papery, necrotic and ultimately die.
- **Manganese:** Deficiency of manganese appears as small chlorotic spots in the interveinal area at the basal part of the leaves and later extends towards the tip. With time these spots turn light greyish-yellow to greyish brown. Under acute deficiency these spots enlarge and coalesce to form streaks of pinkish brown/ buff colour in the interveinal areas.
- **Sodium:** Toxicity symptoms are manifested first on the lower leaves as chlorotic blotches with necrotic lesions starting from the tip and margins and extending inwards. Later the affected leaves wither and dry, with time the whole plant gets affected and gives burning appearance.
- **Boron:** Toxicity first appears on the lower most leaflets. It is manifested as yellowish of tips and margins of leaflets followed by the development of brown to dark brown necrotic lesions giving a burning look to the affected leaves.
- **Copper:** Copper toxicity in rice is exhibited as general chlorosis, resembling iron deficiency. Coupled with drying of leaf tips, stunted growth and reduced tillering are the dominant effects of copper toxicity in rice crop. In case of severe toxicity, the leaves start drying from tip towards base.
- **Nickel:** Nickel toxicity is manifested as general chlorosis of leaves and it resembles iron deficiency. The affected plants show highly stunted growth, reduced branching and root growth is seriously restricted.
- **Cadmium:** It appears as light whitish yellow spots in the interveinal areas of lower leaves. These spots later coalesce and turn yellowish brown. Sometimes leaf tips may also show marginal scorching.
- **Selenium:** Selenium toxicity appears on the young leaves as thin white strips along the midrib which remains green. Later these strips turn brown and necrotic leading to withering of the leaf from tip towards base.

25. FATE OF PLANT NUTRIENT

The plant nutrients released in soil either by weathering or by decomposition of organic materials or by addition of fertilizers are lost by crop removal, leaching, erosion, volatilization, dinitrification, fixation or by combination of any two or more of these processes. The different processes of loss of plant nutrients are discussed below:

1. **Crop removal:** The plant nutrients to a greater extent are lost through various crops grown in the soil. Grain crops exert a greater burden on the soil than

legumes because legume crops utilize a large amount of nitrogen which it fixes. Fortunately many crops require that only the seeds or fruits be removed and much of the other plant parts are returned to the soil. The loss of nutrients are balanced by addition of fertilizers. When there are nutrients imbalance in soil, plants are subjected to nutrient stress that results in reducing growth, yield and quality of crops. Plant differs from each other in respect of nutrient uptake. Legume obtained most of the nitrogen from atmosphere. Plants having extensive proliferated root system (grasses etc) are most competitive for nutrient than less dense root system (potato, beans etc.). Grasses are more competitive for potassium than are clovers. Potato, banana and sugarcane required high potassium. Succulent plants need more nitrogen than seed producing plants like rice, wheat etc. Crop varieties and plant population have variable nutrient requirement.

2. **Leaching:** Water dissolves plant nutrients which are in available form and carries away by percolating down to soil profile or by runoff. Nitrate nitrogen, urea, calcium and magnesium sulphate which are highly mobile are lost through leaching if heavy rainfall or rain follows. About 99 per cent nitrate nitrogen is lost through leaching. In humid region nutrient loss is higher than arid region. Sandy soil having higher percolation rate than loam or clay, loss of much of plant nutrients by leaching. Soils with plant cover permit much less water percolation than barren soils. Intensive cropping minimizes water percolation and reduces the erosion hazards, leaching and surface runoff of valuable nutrients. Potassium and phosphorus are relatively less loss through leaching than nitrogen. The loss of plant nutrients through leaching is influenced by type of soil, climate, amount of rainfall, type of crops and cropping system, presence of vegetation, density of weeds etc.

3. **Volatilization:** Nutrient losses from the surface in the form of gases is called volatilization. Nitrogen is very sensitive and easily lost through volatilization, particularly in rice soils or places when drainage is poor. In alkali or high lime content soil, losses of nitrogen in the form of ammonia is much higher than normal soils. After application of urea under adverse conditions, the losses of nitrogen take place in the following way

$$\underset{\text{(Urea)}}{NH_2CONH_2} + 2H_2O \longrightarrow \underset{\text{(Ammonium carbonate)}}{(NH_4)_2\,CO_3}$$

$$CO(NH_3)_2 + H_2O \longrightarrow (NH_4)_2\,CO_3$$

$$(NH_4)_2\,CO_3 + 2H_2O \longrightarrow \underset{\text{Ammonium Hydroxide}}{2NH_4OH} + \underset{\text{Carbonic acid}}{H_2CO_3}$$

$$NH_4OH \longrightarrow \underset{\text{Ammonia}}{NH_3} + H_2O$$

In anaerobic condition many soil microbes utilize instead of oxygen as electron acceptor the nitrate and removes gaseous nitrogen from soil. The ammonium nitrogen is also lost from soil and escapes as ammonia. As much as 50% of solid ammonium type nitrogen fertilizer has been found to be volatilized when added to surface soil or into irrigation water or to soils of pH about 6.5 or above. Ammonium fertilizers should be pre-plant incorporated to prevent losses from volatilization specially in poor drained soils.

4. **Fixation:** A large amount of available plant nutrients are lost by fixation, precipitation or conversion to a form that is unutilizable which are not available to the plants. Phosphorus is converted from soluble to insoluble form in both acidic and alkaline soils. In acidic soils it is fixed as aluminium and iron phosphate while in alkaline soil as calcium phosphate. Formation of these insoluble forms greatly reduce the phosphorus availability to plants. Phosphorus fixation can be taken place in the presence of calcium and magnesium compounds. In the presence of calcium carbonate, soluble phosphorus converted in unavailable carbonate apatite which is unavailable to the plants. The process can be taken place in the following reaction.

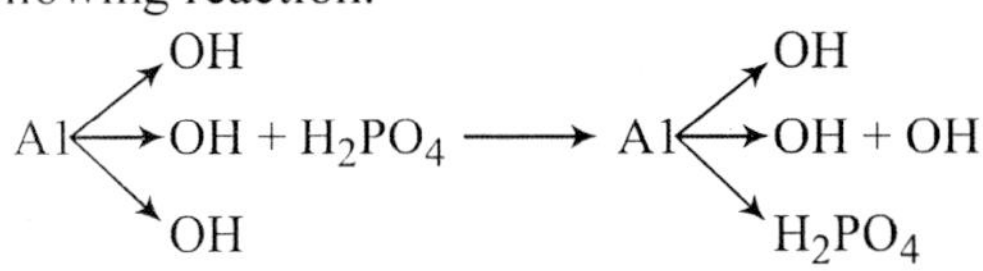

Soluble Insoluble

$$Ca(H_2PO_4)_2 + 2CaCO_3 \longrightarrow Ca_3(PO_4)_2 . CaCO_3 + CO_2 + 2H_2O$$

Soluble Insoluble

If phosphorus is applied in the form of monocalcium, the low amount of calcium reduces the availability of phosphorus. In the process of negative exchangeable, soluble phosphorus fixed on the soil colloids upto that period when Tartrate, Citrate, Oxalate, Fluoride and Hydroxite ions not released phosphorus from the soil colloids. In acidic soils, phosphorus fixation take place in silicate minerals like Kaolinite, Montmorillonite and Illite.

$$[Al] + H_2PO_4 + H_2O \longrightarrow Al(OH)_2H_2PO_4 + 2H$$

Silicate crystal Insoluble

Potassium and ammonium nitrogen can also be fixed in soil by Illite clay minerals which enter into the clay lattice when they swell and are trapped between the lattice on drying. Potassium present in the soil colloids in to three forms such as water soluble potassium, exchangeable potassium and fixed potassium. Water soluble and exchangeable potassium available to the plants while fixed potassium non available form. Under humid conditions potassium fixed in silicate clay like 52 per cent in Illite, 13 per cent in Montmorillonite and 8 per cent in Kaolinite.

5. **Dinitrification:** Conversion of nitrate nitrogen into ammonia or free nitrogen is called denitrification. In the presence of dentrifying bacteria, soil organic matter converted into free ammonia which is lost. Under anaerobic conditions microbial denitrification can take place. Under acid soil conditions denitrification process

occurs in the presence of nitrate, urea, amions and other salts. In silt loam soils of UP and Bihar and black cotton soils of MP denitrification, occurs commonly.

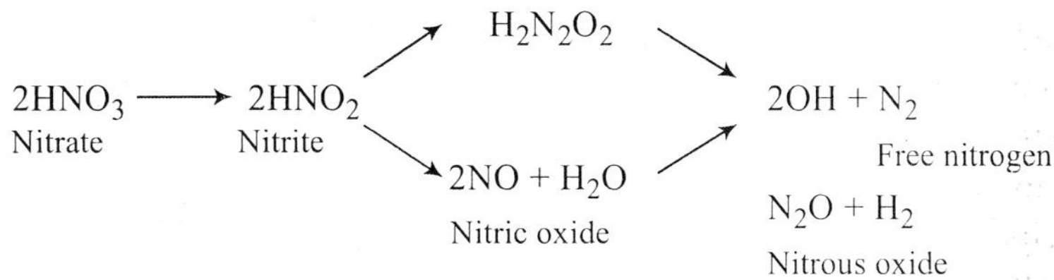

Conversion of nitrate into ammonia in the presence of enzymes.

$$\underset{\text{Nitrate}}{NO_3} \longrightarrow \underset{\text{Nitrite}}{NO_2} \longrightarrow \underset{\text{Hyponitrte}}{HNO} \longrightarrow \underset{\text{Hydroxyl amion}}{NH_2OH} \longrightarrow \underset{\text{Ammonia}}{NH_3}$$

Microbial immobilization or conversion of soluble nutrients to non – available forms removes nutrients especially N, P, K, S from soil atleast for that time when needed by plant most.

6. **Nutrient reservoir:** Nutrient reservoir differs in their capacity to supply nutrient e.g. potassium containing minerals like K schoenite supply less potassium than even sand containing mica. Soils having higher available nutrient supply less nutrients due to high leaching or conversion quickly to non-available form.
7. **Soil erosion:** Large amount of plant nutrients from the upper surface of soil are lost through erosion processes. The extent of plant nutrients loss is dependent on type and amount of erosion. Commonly losses of plant nutrient by water erosion are much more than wind erosion.
8. **By weeds:** Weeds deplete large amount of plant nutrients in the soil. 7-20 % losses occur by weeds. 20-30 kg/ha nitrogen depleted by different weeds in wheat.
9. **Use of microbes:** Presence of bacteria, fungi, microorganisms etc. utilizes plant nutrients for completion of lifecycle. The limited amount of loss of plant nutrients by microorganisms take place in soil.

Exercise 1: Grain and straw samples of rice contained 1.45 % and 0.21% N, respectively. If the grain yield was 67 q/ha and straw yield 71 q/ha, Calculate the total Kjeldahl N taken up by the rice crop.

Since, 100 kg grain contains = 1.45 kg N

Therefore, 67 q grain contains = 1.45 x 67 x 100/100 = 97.15 kg N

Since, 100 kg straw contains = 0.21 kg N

Therefore, 71 q straw contains = 0.21 x 71 x 100/100 = 14.91 kg N

The total N removal by rice crop = 97.15 + 14.91 = 112.06 kg N/ha

Exercise 2: Grain and straw samples of wheat crop contained 0.21% and 0.031% P, respectively. If the grain yield was 55 q/ha and straw yield 65 q/ha, calculate the amount of P removed by the wheat crop in terms of P and kg P_2O_5/ha.

Since, 100 kg grain contains = 0.21 kg P

Therefore, 55 q grain of wheat contains = 0.21 × 55 × 100/100 = 11.55 kg P

Since, 100 kg straw contains = 0.031 kg P

Therefore, 65 q straw contains = 0.031 × 65 × 100/100 = 2.015 kg P

The total P removal by wheat crop = 11.55 + 2.015 = 13.565 kg P

Since, 1 kg P contains = 2.29 kg P_2O_5

Therefore, total amount of P_2O_5 removed by wheat = 13.565 × 2.29 = 31.06 kg P_2O_5

Exercise 3: A barley crop at harvest was separated into grain and straw. The grain contained 0.51% K and straw contained 1.22% K. If the grain and straw yield of barley was 51 and 70 q/ha. Calculate the amount of K removed by the wheat crop in terms of K and kg K_2O/ha.

Since, 100 kg barley grain contains = 0.51 kg K

Therefore, 1 kg barley straw contains = 0.51/100 = 0.0051 kg K

Therefore, 5.1 × 100 × 10 kg barley grain contains = 0.0051 × 5.1 × 100 × 10 /100 = 26.01 kg K

Since, 100 kg barley straw contains = 1.22 kg K

Therefore, 1 kg barley contains = 1.22/100 = 0.0122 kg K

Therefore, 7.0 × 100 × 10 kg barley grain contains = 0.0122 × 7.0 × 100 × 10 /100 = 85.4 kg K

Total amount of K removed by barley = 26.01 + 85.4 = 111.41 kg K

Since, 1 kg K contains = 1.2 kg K_2O

Therefore, total amount of K_2O removed by barley = 111.41 × 1.2 = 133.69 kg K_2O

Similarly other nutrients removal can be worked out.

26. SOIL EROSION

Soil erosion leads to deprivation of physical characteristics of soils and damages plant and crops. In India almost 130 million hectares of land, that is, 45% of total geographical surface area, is affected by serious soil erosion through gorge and gully, shifting cultivation, cultivated wastelands, sandy areas, deserts and water logging.

The removal of soil by running water and wind is known as soil erosion.

Types of soil erosion: Wind and water are the powerful agents of soil erosion because of their ability to remove soil and transport it.

Erosion may be broadly classified into two groups:

1. **Geological erosion:** Otherwise called natural erosion (or) normal erosion represents erosion where land is in natural disturbed environmental condition under the cover of vegetation. Geological erosion takes place as a result of action of water, wind, gravity and glaciers. The soil loss in geological erosion is compensated by the soil forming process by wind weathering.

2. **Accelerated erosion:** When vegetation is removed and the land is put under cultivation the soil gets disturbed, that is the natural equilibrium between soil binding is disturbed.
   (a) **Water erosion:** Loss of soil from land surface by water including runoff from melted snow and ice.
   (b) **Wind erosion:** Erosion of soil by the action of wind is called wind erosion.

Erosion by water: Erosion by water is of several types:

(a) **Sheet erosion:** When a layer of the soil on the surface is removed over a large area by running water, it is called sheet erosion. It remove finer and more fertile top soil.

(b) **Rill erosion:** This is the second stage of sheet erosion. Small finger like rills begin to appear on the landscape. Over a period of time, the fine rills increase in number and also become deeper and wider. This reduces the actual area under cultivation and the yield of crops decline.

(c) **Gully erosion:** When soil is removed by water flowing along definite paths down the slope or in channels, it is called gully erosion. Gullies cut up agricultural land and make it unfit for cultivation.

(d) **Steam Bank erosion:** The continuously flowing water erodes the banks of streams and rivers. Gradually the bed of the river widens.

(e) **Shore erosion:** The tidal waves dash against coastal rocks, causing them to erode bit by bit.

(f) **Slip erosion:** During heavy rains, water percolates into the soil until it is unable to penetrate further by the underlying impervious rocks. On steep land, the heavy moisture – laden soil often comes down bodily, resulting in a landslide.

(g) **Land slides:** Mass movement of sediment from hill sites due to gullies is called landslides.

(h) **Ravines:** Prolonged process of gully erosion leads to ravines and found predominantly is deep alluvial soils.

Specialized forms of water erosion

1. **Pedestal erosion:** Results when a stone (or) the roots protect the soil from the splash erosion, which removes the surrounding soil. It develops slowly over years and is often found on patches on grazing land.
2. **Pinnacle:** Occurs often in gullies as a result of deep vertical rills until pinnacles are left in the land.
3. **Piping:** Piping is the formulation of pipes (or) channels under the soil cover. It occurs when surface water infiltrates through the soil and moves down cards until it covers to a less personable layer.
4. **Slumping:** Slising (or) mass movement of soil in the gully is called slumping. It is usually a process of geological erosion. It is prominent in high rainfall areas with deep soil. Riverbank collapse and coastal erosion also cause slumping.

Soil loss equation: Based on the mechanisms and factors influencing soil erosion, a universal soil loss equation due to water erosion is as below.

$$A = RKLSCP$$

where,

A = Predicted soil loss (ton/acre/yr)

R = Rainfall and runoff factor

K = Soil erodibility

L = Slope length

S = Slope gradient

C = Soil cover and management

P = Erosion control profile

Erosion by wind: Wind erosion is significant in desert and semi-desert regions. In regions with heavy rainfall and steep slopes, erosion by running water is more significant.

Causes of soil erosion

1. **Topography and slope of the land:** In Hilly region, the rainfall occurs in torrents and washes away the top layer of the soil. The steep slopes stimulate the eroding power of rain water.
2. **Nature of rainfall:** Floods and torrential rains cause more damage than light or moderate showers spread over many days.
3. **Nature of soil:** Porous soils with good water – absorbing capacity are least subject to erosion, while the impervious soils are gradually eroded by the action of water.
4. **Human factors:** Man and his activities are responsible for soil erosion to a great extent.
   (a) **Deforestation:** is the removal of trees and other vegetation which hold the soil together. When vegetation is removed, the soil surface is loose and more easily removed by running water and wind.
   (b) **Overgrazing of land:** by animals results in removal of grass over a large area making it easy for wind and running water to remove the soil.
   (c) **Improper farming techniques:** ploughing fields in the traditional up and down manner along the slopes makes it easier for running water and wind to cause erosion.

Consequences of Soil erosion: There are following serious effects of soil erosion which are discussed as below:

(i) Loss of soil stability and fertility.
(ii) Silting up of river bed, lakes and dams.
(iii) Damage of cultivable land and its crops.
(iv) Damage of buildings, roads etc., even erosion takes human life.
(v) Creates flood and cause threat to road communication etc.

Causes of Soil erosion

(i) Bad tillage practices of crop land. Up and down cultivation is to be avoided.
(ii) Over grazing
(iii) Indiscriminate cutting, felling and removal of natural vegetation
(iv) Social and economic factors
(v) Population
(vi) Cattle population
(vii) Small land holding
(viii) Industrializations and development

Control measures

Agronomic practices are:

(i) Contour farming
(ii) Strip cropping— Contour Field, Wind Buffer
(iii) Cover cropping
(iv) Mulching
(v) Crop rotation, sowing of leguminous crops and mixed cropping
(vi) Tillage

Mechanical method/Engineering measures/Structural measures may be broadly classified as:

1. Erosion control measures

(i) Bunding
- a. Contour bunding
- b. Graded bunding

(ii) Bench terracing
- a. Level terrace
- b. Inward sloping terrace
- c. Out ward slopping terrace

(iii)Trenching
- a. Contour trencing
- b. Graded trenching

2. Drainage Line treatment measures

(a) Check dam.
- (i) Temporary check dam
- (ii) Semi permanent structures
- (iii) Permanent gully control structures

(b) Drop structure
(c) Chute spilway
(d) Drop inlet spilway
(e) Gabion check dam
(f) Spurs

27. SOIL REACTION

Soil reaction is one of the most important chemical properties of the soil solution that influences many physical, chemical and biological properties of soil. It is also the most important factor for evaluating its production potential for most crops.

Soil reaction is an indication of the acidity or basicity of the soil and it has been classified into three groups (1) Acidity, (2) Alkalinity, and (3) Neutrality.

pH is defined as the logarithm of the reciprocal (or negative logarithm) of the hydrogen – ion concentration in gram per litre; represented in equation as follows:

$$pH = \log \frac{1}{[H^+]}$$

At neutrality, the hydrogen-ion concentration is : 0.0000001 or 1×10^{-7} gram of hydrogen per litre of solution. Substracting this concentration into the formula,

$$pH = \log \frac{1}{0.0000001}$$

$$pH = \log 10000000 = 7$$

At a pH of 6, there is 0.00001 gm of active hydrogen, or 10 times more the concentration of H^+ increases by 10 in concentration. It therefore follows that a pH of 6 is 10 times more acidic than a pH of 7; a pH of 5 is 10 times more acid than a pH of 6, and so on.

- The pH value, therefore, represents the amount of free or active acidity and not the total quantity of potential (or combined) acidity.
- It represents the intensity of acidity of a solution.
- It represents concentration or activity of hydrogen ion.
- pH expresses the relationship between H^+ and OH^-. Alkalinity is due to excess of OH ions over H ions.
- A neutral reaction is produced by an equal activity of H and OH ions.
- The pH value ranges from 0-14, where pH 0 represents the highest limit of active acidity, and pH 14 the highest degree of basicity (or alkalinity). Neutrality represents pH 7. Therefore, pH 7 shows a neutral reaction. When the pH is less than 7, the solution is acidic, when it is above 7, it is alkaline.

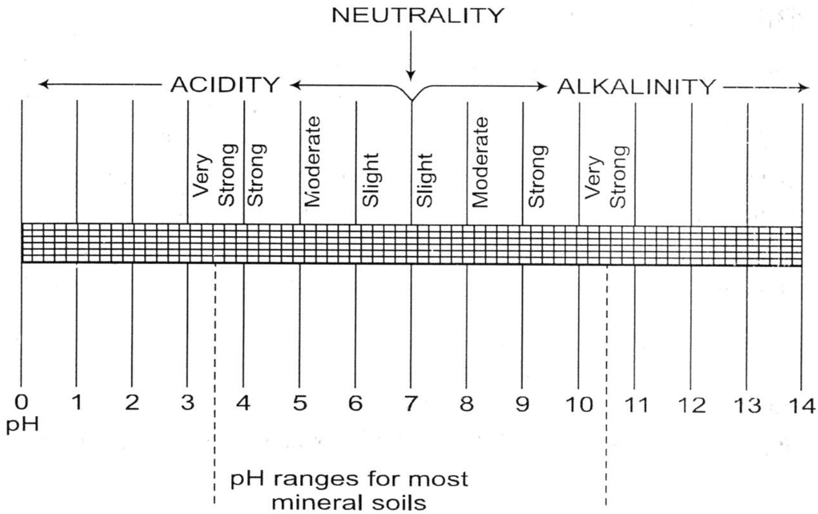

Ranges in pH

Relationship between H^+ and pH

H^+	pH
1×10^{-0}	0
1×10^{-1}	1
1×10^{-2}	2
1×10^{-3}	3
1×10^{-4}	4
1×10^{-5}	5
1×10^{-6}	6
1×10^{-7}	7
1×10^{-8}	8
1×10^{-9}	9
1×10^{-10}	10
1×10^{-11}	11
1×10^{-12}	12
1×10^{-13}	13
1×10^{-14}	14

Relationship between pH and pOH

pH	Acidity (Normality of H^+)	Alkalinity (Normality of OH^-)	pOH
0	1.0	0.000 000 000 00001	14
1	0.1	0.000 000 000 0001	13
2	0.01	0.000 000 000 001	12
3	0.001	0.000 000 000 01	11
4	0.0001	0.000 000 000 1	10
5	0.00001	0.000 000 001	9
6	0.000001	0.000 000 01	8
7	0.0000001	0.000 0001	7
8	0.00000001	0.000001	6
9	0.000000001	0.00001	5
10	0.0000000001	0.0001	4
11	0.00000000001	0.001	3
12	0.000000000001	0.01	2
13	0.0000000000001	0.1	1
14	0.00000000000001	1.0	0

Factors controlling soil reactions

1. **Nature of soil colloids:** When hydrogen (H^+) ion forms the predominant adsorbed cations on clay colloids, the soil reaction becomes acid.
2. **Soil solution:** The increases in the acidity of the soil solution lowers its pH. The soil solution carries a number of salts dissolved in capillary water.
3. **Climate:** In general soil formed in region of high rainfall are acidic (low pH), while those formed in regions of low rainfall are alkaline (high pH).
4. **Soil management:** Cultural operations in general tend to increase soil acidity. They make an acid soil more acidic, and an alkaline soil less alkaline.
5. **Microbial action:** Decomposition of organic matter produces carbonic acid (H_2CO_3). Sulphuric and nitric acids are formed by the organic decay processes and by microbial action on inorganic sulphur and nitrogen containing materials. These acids encourage acidic conditions in the soil.
6. **Formation of sulphuric and nitric acids in the atmosphere.**The precipitation of these substances is called acid rain. The pH value of acid rain may range from 2.0 to 4.5.

Influence of soil reaction on availability of nutrients:

1. **Nitrogen availability:** Nitrification is highest where the pH is between 6.5 to 7.5. It is adversely affected if the pH falls below 5.5 and is greater than 9.0. Nitrogen fixing bacteria like Azatobactor also fails to function below pH 6.0.

The decomposition of organic matter which is primary source of nitrogen also slow down under acidic conditions.

2. **Phosphorus availability:** Its availability is highest when pH is between 6.5 to 7.5. Fixation of P takes place in strongly acid soil:

$$\underset{\text{(Soluble)}}{Al^{+++} + H_2PO_4^- + 2H_2O} \rightleftharpoons \underset{\text{(Insoluble)}}{2H^+ + Al(OH)_2.H_2PO_4}$$

The phosphates react with hydrated oxides of iron and aluminium and form is soluble hydroxy phosphates of iron and aluminium. Unavailability of phosphorus is called phosphorus fixation.

Fixation of P takes place even when the soil is alkaline or has high pH:

(i) $\underset{\text{(Soluble)}}{(Ca(H_2PO_4)_2 + 2Ca^{++}} \longleftrightarrow \underset{\text{(Insoluble)}}{Ca_3\ (PO_4)_2} + 4H^+$

(ii) $\underset{\text{(Soluble)}}{Ca(H_2PO_4)_2} + 2CaCO_3 \rightleftharpoons \underset{\text{(Insoluble)}}{Ca_3(PO_4)_2} + 2CO_2 + 2H_2O$

The availability of phosphorus at different pH is linked with the ionic form in which it is present in soil solution. The monovalent $H_2PO_4^-$ ions predominate in highly acid solution with decreasing acidity, the divalent HPO_4^{--} ions begin to appear. In alkaline soil the trivalent PO_4^{--} ions are present in extremely small quantity. At pH 9.0 and above, PO_4^{--} is available to plants.

In acid soils, phosphorus becomes available by anion exchange. Part of the phosphate which has react with iron and aluminium compounds is subject to replacement by other ions such as the hydroxyl ion. Such replacement is called anion exchange.

$$\underset{\text{(Insoluble)}}{Al(OH)_2 \cdot H_2PO_4} + OH^- \rightleftharpoons Al\ (OH)_2 + \underset{\text{(Soluble)}}{H_2PO_4}$$

One anion (OH) has been exchanged for another (H_2PO_4). Phosphorus ($H_2PO_4^-$) become available after liming in the acid soils.

3. **Potassium:** In acid soil, K is lost through leaching whereas in alkaline soil, solubility of K is depressed results in non-availability. The unavailability of K is due to the conversion of exchangeable to non-exchangeable potassium.
4. **Calcium and Magnesium:** Acid soils are poor in available Ca and Mg whereas Ca and Mg are available more in alkaline soil (pH not excedding 8.5). When the pH is above 8.5, the availability of these nutrients again decreases.
5. **Iron, Aluminium and Manganese:** When the pH is low, the solubility of iron, aluminium and manganese compounds is increased. Under neutral and alkaline conditions, Fe and Mn are usually available.
6. **Sulphur:** The availability of S is not affected by soil reaction. It is more soluble in acid soil and lost in leaching. The availability of sulphur present in organic matter depends open the decomposition of organic matter.
7. **Micronutrients:** In general, the availability of boron, copper and zinc is reduced in alkaline soil and that of molybdenum in acid soil. The availability

of boron, coppers and zinc progressively decreases as the soil pH increases. Their availability also decreases under highly acid condition. In alkaline soils soluble zinc converted into insoluble zinc salts (calcium zincate) which reduces its availability. Zinc and copper adsorbed on the clay colloids are not easily displaced and hence, not available for plant growth. The availability of molybdenum is reduced under acid soils. It is more available in neutral and alkaline soils.

Buffering and soil reaction: Buffering refers to resistance to a change in pH. The colloidal complex thus acts as a powerful buffer in the soil. Buffering regulates the availability of plant nutrients. Buffering prevents sudden changes and fluctuations in soil pH, so it regulates the availability of nutrients and also checks direct toxic effects to plants. Buffer tends to ensure reasonable stability in the soil pH and prevents drastic fluctuations that might be detrimental to microorganism and plants. Buffer also influences the amount of chemicals, such as lime application in acid soil.

Buffering capacity

The resistance of soil to changes in pH as a result of the addition of acid or basic material is called buffering capacity of the soil. Pure sand has almost no buffering ability, while soils high in colloids have high buffering capacity. The general principle is that an increase in a particular cation in the soil solution will cause that cation to be fixed to colloide (buffered) and decrease in solution of that cation will cause it to be withdrawn from the colloid and moved into solution (buffered). The degree of buffering is limited by the CEC of the soils; the greater the CEC, the greater the buffering capacity of the soil.

Soil reaction and plant growth

Most of the plants can grow on soils with a wide range of pH, However, plants have preference for a specific range of soil pH. Soil reaction and plant growth are interrelated. It increases or decreases the availability of nutrient elements. For example, in acid soil the concentration of Al^{+++}, Fe^{+++} or Mn^{++} increases to the toxic level. High pH generally lowers availability of the nutrient elements except molybdenum and boron.

Soil reaction (pH) preference of crop plants

Crops	Optimum pH range
Cereals	
Rice	4.0 – 6.0
Wheat	6.0 – 7.5
Maize	6.0 – 7.5
Millets	5.0 – 6.6
Barley	6.0 – 7.5
Oat	5.0 – 7.5

Legumes	
Groundnut	5.3 – 6.5
Pea	5.5 – 7.0
Lentil, gram etc.	5.5 – 7.0
Soybean	5.5 – 7.0
Berseem	6.0 – 7 5
Other field crops	
Potato	5.5 – 6.5
Cotton	5.0 – 6.5
Sugareane	6.0 – 7.5
Tea	4.0 – 6.0

28. SOILS OF INDIA

Soils in India have been classified into a large number of groups and subgroups. The main groups are as follows:

1. Red soils
2. Black soils
3. Laterite soils
4. Alluvial soils
5. Hills soils
6. Desert soils
7. Saline and alkali soils
8. Peaty soils
9. Tarai soils

1. **Red soils:** Red soils generally found in whole Tamilnadu, part of AP, MP, Orissa, Bihar, UP (Bundelkhand), WB (Birbhumi)) and Rajasthan. Red soils are further divided into red loam and red sandy soils.

(a) **Red loam:** These soils are derived from the weathering of granites, gneisses, schists, dionites etc, which are relatively rich in clay forming minerals and poorer in silica. The red colour of soil is due to the presence of various oxides of iron. The clay mineral is mainly kaolinite. Finer texture soils are formed by the base rich minerals like, feldspar, mica, hornblende etc. They are poor in fertility, low base exchange capacity, deficient in organic matter and low water holding capacity. Crop can be raised under protective irrigation or on conserved moisture situation. Monocropping is predominantly practised in these soils. The pH is around neutral or slightly acidic. Some soils have high pH. Depth of surface horizon is 25 to 40 cm and B horizon is 0 to 80 cm.

(b) **Red sandy soils:**These soils are derived from granites, graniloid, gneisses, quartzites and sandstone. The colours given by red haematite and/or yellow limonite. Clay minerals are mainly Kaolinite and Illitic types with smaller amounts of montmorillonite. Base exchange capacity is from 5 to 25 meq per 100 gm of soil and pH generally towards the acidic. A moderately deep A horizon and B horizon of 15-30 cm.

2. **Black soils:** The black soils are found in parts of Maharastra, Gujarat, MP, Rajasthan, UP, AP, Tamilnadu and Karnataka. They are dark in colour, varying from dark brown to deep black. The composition of clay is chiefly of the montmorillonite group and thus swelling and shrinking type. The pH varies from 7.5 to 8.5. They are formed by the weathering of trap rocks. These soils are classified into three groups:
   (a) **Shallow black soils** are derived from basalts of Deccan traps. The texture varies from silt loam to clay with a depth of 30 cm. The colour of these soils varies from dark brown to dark yellowish brown. The structure is granular or weakly blocky. Lime is usually present in fine grains or nodules.
   (b) **Medium deep black soils** develops from a variety of rocks including basaltic traps, schists, basic granites, gneisses, hornblende and chloritic schists. Texture varies from silty clay to clay with a depth of 30 to 100 cm. These soils are moderately rich in organic matter and are fairly well drained.
   (c) **Deep black soils** are derived from basaltic traps. The percentage of clay may vary from 40 to 60. The clay mineral is of 2:1 latice type structure. Lime is present in irregular shape and therefore, soil reaction is generally alkaline.
3. **Lateritic soils:** These soils are found in Karnataka, Kerala, Maharashtra, Orissa, MP and UP. Yellow-red or red colour soils derived from laterites and contain a large proportion of primary kaolinite clay minerals. These soils exhibit plasticity, cohesion, shrinkage, expansion and base saturation qualities to a small extent. The pH of these soil is generally on the acidic side (5.0 – 6.0) due to the lack of lime and magnesium. Due to the breaking off and forming again in different places two types of laterites may be found, low level laterites and high level laterites. They are poor in water retention. The soils are fairly good in organic matter content but deficient in phosphorus and potassium. The base exchange capacity of mineral colloids may range from 2 to 4 meq/100 gm of soil.
4. **Alluvial soil:** These soils cover large part of Rajasthan, Gujarat, Punjab, Haryana, UP, Bihar, WB and some part of Assam and Orissa. These soils are developed mainly by the depositions of silt over ages. The older alluvium is more clayey and is darker in colour with high lime content. The fresh alluvium are coarse in texture and show little or no horizonation. Considerable salinity and/or alkalinity problem is also found. Alluvial soils are derived into four categories:
   (a) **Coastal alluvium:** The texture is extremely variable and ranges from sandy to silty clay. The soils are usually deep and ranges from bright reddish brown to yellowish brown and to grey and dark grey. Soils derived from calcareous material are composed of dark coloured heavy clay. If these soils are derived from red soils which are derived from granite and gneissic rocks, the soils are poor in fertility with medium or light texture and rich in kaolinite clays.
   (b) **Coastal sands:** The soils are sandy and deep but lack in profile development. Due to low water table and free water drainage, the salinity is not a problem. These sandy structures are put under coconut and cashew plantation.

(c) **Deltaic alluvium:** These soils are brought by rivers and deposited where they join the sea. These soils vary in their composition and texture, depending on where they are form. The drainage condition is most of the soil satisfactory, the exception being Gangetic Alluvium.

(d) **Calcareous Alluvial soils:** These calcareous soils developed on the alluvium of the Gandak river. The main characteristics being a high content of calcium carbonate, ranging from 10 to 40%, and distributed throughout the profile. The soils are light coloured pale brown and yellow brown and lack horizon differentiation. The texture varies from sandy loam to loam and pH is mainly alkaline. Phosphorus and potash content are low.

5. **Desert soils:** Desert soils are commonly found in Rajasthan, South Punjab and in Kutchh. The most predominant part of desert soils is quartz but feldspar and hornblende, grains also occur. These are two distinct types of soils:

(a) **Regosolic:** The regosol desert soils consists of wind blown sand and sandy fluviatile deposits. The depth of these soils extend over 50 cm. The A and B horizons are weakly developed. They are low in organic matter content and colour varies from yellowish brown to pale brown and contains many weatherable minerals. The pH is fairly high with high amount of calcium carbonate and soluble salts.

(b) **Lithosolic:** Colour of these soils are rusty brown and dark reddish brown, moisture retentivity is low and no plant life is sustained.

6. **Tarai soils:** These soils are mainly found in foot hills of Himalaya, J&K, UP, Bihar and WB. They are formed by the downward movement of materials from the lower Himalayan range. Surface soil possess a sandy loam to silty loam texture with adequate drainage. These are fertile soils.

7. **Hill soils:** These soils are commonly found in hilly region of Himalaya. They are formed from sandstone, grey mica, sand stones and shales. The surface of these soils are dark brown and loam to silty clay and moderately rich in organic matter. They have a high base exchange capacity owing to the high organic matter content. They lack in free lime and have pH near to neutral or on the acid side.

8. **Saline and alkali soils:** The Indo-Gangetic alluvium in the North and deltaic, coastal region, south and west soil region have these soils. These soils are characterized by the presence of high concentrations of soluble salts of alkali base. Introduction of irrigation in certain areas have changed the soil to alkali (saline) soils. The salts rise to the surface with the introduction of irrigation as in the case of black soil and salts accumulate. Saline soils are considered better than alkali soils as the soil is not dispersed and can be rectified with irrigation and the application of sulphur or gypsum.

9. **Peaty soil:** These soils are found in Kerala, North Bihar and North UP. They develop from brackish water sediments containing appreciable amounts of pyrites. By oxidation and hydrolysis of sulphur compounds, sulphuric acid is developed and the pH goes down to 3 or 4. These soils contain a high percentage

of free alumina and iron due to the action of sulphuric acid. These have higher accumulation of organic matter due to poor drainage conditions.

29. CLASSIFICATION OF SOIL GROUPS

According to new classification soil groups are classified as given below:

Sl. No.	Soil classification	Area (Sq. km)	Distribution (State)	Classification
1.	Shallow black soils	31532	Maharashtra	Ustrothents Ustopepts
2.	Medium black soils	430303	MH, MP, Gujarat	Pellusterts Chromusterts
3.	Deep black soil	112060	MH, AP, Karnataka, MP, Gujarat	Pellusterts Chromusterts Pelluderts
4.	Red sandy soil	330590	TN, KN, UP	Haplustalfs Rhodustalfs
5.	Red loamy soil	213271	AP, TN, KN, MP, Kerala	Paleustalfs Rhodustalfs Haplustalfs
6.	Laterite soil	130056	MH, Orissa, AP, TN, Kerala, Goa, Assam	Plinthuaqulta Plinthustults Plinthudults Oxisols
7.	Red and yellow soil	403651	MP, Orissa	Haplustults Ochroqults Rhodustults
8.	Mixed red and black soils	162255	MP, MH, KN, TN	Alfisols Vertisols
9.	Alluvial soil Khadar (New alluvial) Bangar (Old alluvial)	355720	UP, Punjab, Bihar, WB, Assam	Haplaquants Ustifluvents Udifluvents Heplustalfs Ustocrepts
10.	Alluvial soil (High calcium)	13611	North-East UP, Bihar	Calciorthents
11.	Deltaic alluvium soil	87045	WB, Orissa, AP, TN	Quartzipsamments Tropaqualfs
12.	Coastal alluvium soils	54403	MH, Gujarat, AP, TN, Kerala	Haplaquents
13.	Coastal sandy soils	5534	Orissa	Ustipsamments

14.	Lime sierozemic soils	45080	Punjab	Calciorthids
15.	Desert soils	154423	Rajasthan	Calciorthids Psamments
16.	Grey brown soils	101572	Rajasthan, Gujarat	Calciorthids
17.	Tarai soil	28919	UP, Bihar, WB	Heplaquolts
18.	Lithosolic soils	-	-	Lithic, Entisols
19.	Brown hills soils	81242	UP, Sikkim, HP	Palehyumults
20.	Sub mountain soils	76695	UP, J&K	Hepludalfs
21.	Mountain meadow soils	59790	Kashmir and Ladhakh	Cryoborols Cryochrepts
22.	Saline and Alkali soils	173777	Gujarat, WB, Rajasthan, UP, Pb, MH, KN, TN	Solar thids Salorgids Natargids, Natric
23.	Peaty soils	2720	Kerala	Histosols
24.	Skeletal soils	79151	MP	Lithic Entisols
25.	Glacier soil	29335	Uttrakhand	

30. SALINE AND ALKALI SOILS

Presently 29.60 lakh hectare land is suffering from salinity and alkalinity. Gujarat comprised maximum soil affected soil (16.81 lakh ha) followed by West Bengal (4.412 lakh ha.), Rajasthan (1.956 lakh ha), Maharastra (1.841 lakh ha), Orissa (1.471 lakh ha), Andhra Pradesh (0.776 lakh ha), Haryana (0.492 lakh ha), Bihar (0.473 lakh ha) and Uttar Pradesh (0.220 lakh ha).

Saline and alkali soils are divided into three groups:

- **Saline soils:** These soils are locally named as reh, rehala or Namkin. In this type of soil there is the presence of extensive white greyish, white or ash coloured fluffy deposit of salts on the surface of land. The texture of soil may vary from loamy sand to loam, but the subsoil are not compact. There is usually no hard pan or kankar layer in the subsoil. Such soil usually occurs in the vicinity of the rivers, streams or canals where obstruction to natural drainage has been taken place and consequently ground water table remains near the surface for the major part of the year. The pH of the soil usually remains below 8.5.
- **Alkali soil:** Alkali soils are the typical usar soils. The profile possesses a hard and compact structure with loam to clay loam texture, the surface soil of which rests on heavier indurated and columnar subsoil. When wet, the soils are plastic and gummy. Physical deterioration of the soil is of extreme degree. Usually, a thick and highly cemented bed of big sized kankar is to be found in the subsoil which is occasionally so much cemented that it appears rock-like and difficult to pierce. Water percolation is almost nil and surface water stagnation with high degree of turgidity is a common land feature. On evaporation and drying, such

condition may leave flakes of clay and black redish-brown residue on the surface (black alkali) caused by the dissolution of humus in the alkaline media. The soils are completely barren or else can support only salt – tolerant grasses during the rainy season. Exchangeable sodium percentage is only a little above 15 but the salinity is low and EC is below 4 dsm^{-1}.

Alkali soils are containing sufficient exchangeable sodium to interfere growth of crop plants but these do not contain appreciable quantities of soluble salts. The pH of such soils usually ranges between 8.5 – 10.0.

- **Saline and alkali soils:** Saline – Alkali soils are containing soluble salts in sufficient quantity to interfere with the growth of crop plants and contain sufficient exchangeable sodium which affects the soil properties and plant growth adversely. The pH of saline –alkali soils is higher than 8.5.

  About 7 million hectare of land are affected by soil salinity/alkalinity in India. Locally these soils are known by different name in different parts of the country viz., Reh, Rehala and Usar in Uttar Pradesh, Thur, Kallar, Rakkar, Bara and Bari in Punjab and Haryana, Khar and Lona in Gujarat, Chopan and Khari in Deccan, Choudu and Uppu in Andhra Pradesh and Tamilnadu, Khari in Kerala etc.

Difference between saline and alkali soils

Saline soils	Alkali soils
1 These soils are formed by accumulation of soluble salts in higher concentration	These soils are formed from saline soils by desalinisation and accumulation of hydroxides of sodium and potassium.
2. They are called white alkali due to accumulation of soluble white salts on the surface	They are called black alkali due to accumulation of alkali soluble humus, a dark coloured complex on the surface.
3. The electrical conductivity of their saturation extract is more than 4 dsm^{-1}.	The electrical conductivity of their saturation extract is less than 4 dsm^{-1}.
4. The soils have less than 15 per cent of exchangeable sodium.	The soils have more than 15 per cent of exchangeable sodium.
5. These soils are slightly alkaline and pH remains below 8.5.	The soils strongly alkaline and pH is usually more than 8.5.
6. Soils contain more than 2 % harmful soluble salts.	Soils contain less than 0.2 % harmful soluble salts.
7. The osmotic pressure of the soil solution remains high.	The osmotic pressure is less than saline soils.
8. Soils contain generally higher concentration of chlorides and sulphates of sodium but lesser amount of bicarbonates and carbonates.	Soils contain lesser amounts of chlorides, sulphates but higher concentration of bicarbonates and carbonates.
9. Soils do not contain, sodium carbonates or carbonate anions are absent.	Soils contain soluble sodium or potassium carbonates or carbonate anions are present.
10. The physical conditions of these soils remain favourable for cultivation.	Their physical conditions become very poor and are unfit for cultivation.
11. These soils can easily be reclaimed.	They are difficult to be reclaimed.

Characteristics	Saline soil	Alkali soil	Saline-alkali soil
pH	<8.5	8.5 - 10	>8.5
Electrical conductivity	>4.0	<4.0	>4.0
ESP	<15.0	>15.0	>15.0

Sodium Adsorption Ratio

$$SAR = \frac{Na}{\sqrt{\frac{Ca^{++} + Mg^{++}}{2}}}$$

Residual Sodium Carbonate (RSC) : $(CO_3^{--} + HCO_3^{-}) - (Ca^{++} + Mg^{++})$

Salt concentration (mg/L) = E.C. (dSm^{-1}) at 25 °C × 640

Salt concentration (Me /L) = E.C. (dSm^{-1}) at 25 °C × 10

Reclamation of saline and alkali soils

(A) Chemical amendment: Various chemical amendments suitable for different soil conditions are:

1. **Gypsum:** The chemical formula of gypsum is $CaSO_4.2H_2O$. It reacts with exchangeable sodium which gets converted in to sodium sulphate. Sodium sulphate is leached from the soil to reduce the soil pH. The following chemical reactions takes place:

   $$CaSO_4 + 2Na\text{ - clay} = Ca\text{ - clay} + Na_2\,SO_4$$

   The rate of application of gypsum is usually 2.5 – 5.0 tons per hectare. But it can be safely applied even upto 5 to 10 tons per hectare depending on the soil pH and the concentration of exchangeable sodium present in the soil complex.

   Gypsum reacts with sodium carbonate to convert calcium carbonate and removes sodium as sodium sulphate.

   $$Na_2CO_3 + CaSO_4 = CaCO_3 + Na_2SO_4$$

   Gypsum is a deep and very effective soil amendment for the reclamation of alkali soils. Its amounts differs with pH of soil. The quantity of gypsum to be used for reclamation and saline-alkali soils can be calculated by determining gypsum requirement (G.R.).

2. **Sulphur:** In high calcium containing alkali soils, sulphur dust is applied which is mixed with the surface soil with light irrigation. Sulphur is oxidized by soil microbes in following steps.

   $$2S + 3O_2 \longrightarrow 2SO_3$$

   $$SO_3 + H_2O \longrightarrow H_2SO_4$$

   The sulphuric acid thus produced, removes carbon dioxides and neutralizes the alkalinity.

   $$H_2SO_4 + Na_2CO_3 \longrightarrow Na_2SO_4 + H_2O + CO_2$$

$$H_2SO_4 + CaCO_3 \longrightarrow CaSO_4 + CO_2 + H_2O$$
$$2Na\text{-clay} + H_2SO_4 \longrightarrow 2H\text{-clay} + Na_2SO_4$$

In the absence of carbonates it removes sodium from soil colloids

$$2Na\text{-clay} + H_2SO_4 \longrightarrow 2H\text{-clay} + Na_2SO_4$$

Generally 0.32 ton sulphur is required for exchanging per milli-equivalent sodium from 1 hectare soil.

3. **Pyrite:** Pyrite consists of iron and sulphur minerals (FeS_2) usually found in igneous and metamorphic rocks. It is composed of 22-24 % sulphur, 20-22 % iron, 0.5 – 0.6 % calcium and magnesium oxide, 6-8% aluminium, 35 – 40 % silica, 2-3 % carbon and traces of zinc, copper and manganese. When sulphur is applied to alkali soils, it forms sulphuric acid.

$$FeS_2 \longrightarrow FeS + S$$
$$2FeS \longrightarrow Fe_2O_3 + 2S$$
$$2S + 3O_2 \longrightarrow 2SO_3$$
$$2SO_3 + 2H_2O \longrightarrow 2H_2SO_4$$

Sulphuric acid decomposes carbonates and removes sodium from soil exchange complex.

$$CaCO_3 + H_2SO_4 \longrightarrow CaSO_4 + CO_2 + H_2O$$
$$2Na - clay + CaSO_4 \longrightarrow Ca\text{-clay} + Na_2SO_4$$
$$2Na\text{-clay} + H_2SO_4 \longrightarrow 2H\text{-clay} + Na_2SO_4$$

4. **Sulphuric acid:** Sulphuric acid neutralizes the alkalinity of soil, but being expensive, it is used in limited areas.
5. **Liquid sulphur di-oxide:** Liquid sulphur dioxide gets oxidized in soil and forms sulphuric acid which neutralizes the alkalinity of soil.

$$2SO_2 + O_2 \longrightarrow 2SO_3$$
$$SO_3 + H_2O \longrightarrow H_2SO_4$$

6. **Iron and aluminium sulphate:** Sulphates of iron and aluminium are acidic salts and produce acid through hydrolysis. The acid removes the exchangeable sodium with H ions.

$$FeSO_4 + 2H_2O \longrightarrow \underset{\text{Insoluble}}{Fe(OH)_2} + \underset{\text{Soluble}}{H_2SO_4}$$
$$Al_2(SO_4) + 6H_2O \longrightarrow 2Al(OH)_3 + 3H_2SO_4$$
$$2Na\text{-clay} + H_2SO_4 \longrightarrow 2H\text{-clay} + Na_2SO_4$$
$$CaCO_3 + H_2SO_4 \longrightarrow CaSO_4 + CO_2 + H_2O$$

7. **Lime stone:** Lime stone is used in calcium deficient alkali soils. Calcium of lime stone is exchanged with sodium.

$$2Na\text{-clay} + 2H_2O \longrightarrow 2NaOH + 2H\text{-clay} \quad ...(1)$$
$$2H\text{-clay} + CaCO_3 \longrightarrow Ca\text{-clay} + CO_2 + H_2O \quad ...(2)$$
$$2Na\text{-clay} + \underset{\text{Lime stone}}{CaCO_3} \longrightarrow Ca\text{-clay} + Na_2CO_3$$

8. **Organic amendment:** The organic amendment commonly used are farm yard manure, compost, green manures, molasses, pressmud, vegetative residues, weeds etc. Through microbial decomposition of organic matter produces carbon dioxide which dissolves in soil solution and forms carbonic acid. The carbonic acid solubilises calcium salts and reduces the alkalinity of soil. The organic matter improves physical and chemical properties of soil and permeability of soil.
9. **Growing of salt tolerant crops:** Some crops are very sensitive and cannot tolerate the excess of salts but there are many crops which are tolerant (resistant) to moderate salt concentration.
   (i) **Low salt tolerant crop** - Beans, Green gram, blackgram, Chickpea, Fieldpea, Alfalfa, Oranges, Grapefruit, Berry, Plum, Almond, Straw berry etc.
   (ii) **Moderate salt tolerant crops-** Beans, Rye, Wheat, Barley, Paddy, Millets, Maize, Arhar, Tomato, Cabbage, Cauliflower, Potato, Raddish, Carrot, Onion, Cucurbits, Sweet clover, Bottle guard, Berseem, Cowpea, Oats, Fig, Grapes, Pomegranate, Guava, Palm, Citrus, Strawberry etc.
   (iii) **High salt tolerant crops-** Dhaincha, Sugarbeet, Paragrass, Tobacco, Turnip, Carrot, Gardenpea, Rape, Cotton, Palm, and Rhodes grass etc.

31. ACID SOILS

Soils having pH less than 5.5 are called acid soils. There are two criteria used to distinguish acid soils from non acid soils.

1. Percent Base Saturation (PBS) =

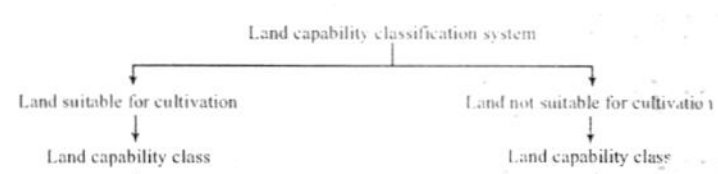

- The unit of PBS are Cmol/kg = m.eq./100g of soil
- Humid soils, dominated by silicate clays and humus are acid if PBS is much below 80%.
- Good cultivated land has 60% PBS occupied by calcium. It indicate that the calcium dominant soils are fertile in nature.
- At a given percent base saturation, the pH is normally highest in hydrous oxide clay minerals, intermediate in kaolinite and lowest in 2:1 type minerals.
- The greatest buffering occurs at about 50% saturation.
- The buffering is uniform over pH range 4.5 – 6.5. It indicates that same amount of lime is required to the soil pH from 5.0 – 5.5 as from 5.5 – 6.0.

2. **pH :** The major source of H^+ in most acid soils is adsorbed aluminium.

$$Al^{3+} + H_2O \longrightarrow Al(OH)^{2+} + H^+$$

- Adsorbed hydrogen is the second but minor source of H^+.
- Acid soils have relatively high amount of exchangeable H^+ and Al^{3+}. Kaolinite and illite clay minerals are dominant.

- Due to continuous rains, the bases obtained by ionic exchange and mineral decomposition, get leached away and insoluble acid residues remain in which Si, Al and Fe oxides are dominant and responsible for acidity.
- Acid soils are rich in quartz and silica. Rocks like granite and rhyolite produce silicic acid mainly orthosilicic and trisilicic acids. When these rocks lack in base, produce acidity in soil after decomposition by weathering.
- Long term use of acid fertilizers like ammonium sulphate.
- Leaching of salts in fallow land causes acidity.
- Potato and beets absorb salts from soil causing increase in the acidity.
- Carbonic acid formation by nitrification and reduction increase acidity.
- The CO_3^{2-} and HCO_3^- after decomposition by chemical process are leached away due to being soluble in nature and hydrated aluminium silicate causes acidity.

Acid soil formation process

- Laterization of varying degrees.
- Podzolization with sub-temperate to temperate climate.
- Intense leaching of light alluvial soils.
- Marshy conditions with significant amount of partly decomposed organic matter.

Adverse impact

Plants:

- Toxic effects of H^+ ions on root tissues.
- Influence of soil acidity on the permeability of the plant membrane for cations.
- Disturbance in the balance between basic and acidic constituents through the roots.
- Harmful effect on enzymatic changes.
- Hampered availability of P, Cu and Zn plant nutrients.
- High solubility and availability of elements like Al, Mn and Fe in toxic amounts.
- Soil micro-organisms are adversely affected.
- More incidence of plant diseases.
- Nutrients such as Ca and K may be deficient.

Soils:

- **Physical conditions:**
  - Light texture
  - High permeability
  - Poor water holding capacity
  - Poor cation exchange capacity
  - Poor organic matter content
  - Deflocculated conditions of soil due to high acidity

- **Chemical conditions:**
  - Base saturated soil, more anions than cations
  - Active and potential soil acidity
  - Low availability of plant nutrients such as Al, Fe, Mn, Zn, Cu, Co
  - High availability of plant nutrients such as Mo, N, K, Ca, Mg, S
  - More Al^{3+} concentration affect toxicity on plants

Management of acid soils: Liming is generally recommended for acid soils.

For normal and 4.8 pH – 35 t/ha lime is required.

For pH 6-7 – 4 t/ha lime is required:

Liming materials:

1. Calcium limestone ($CaCO_3$) is mostly used
2. Dolomite (rich in Mg)
3. Quick lime (CaO)
4. Slaked lime ($Ca(OH)_2$)
5. Coral shell lime
6. Chalk ($CaCO_3$)
7. Blast furnace slag ($CaSiO_3$ and Ca_2SiO_4)
8. Wood ashes

32. LAND CAPABILITY CLASSIFICATION

As per the International Land Capability Classification which has also been adopted in India, lands have been categorized into eight capability classes ranging from I to VIII. Lands belonging to I to IV are suitable for cultivation and those from V to VIII are not suitable for cultivation. Factors like texture, soil depth, permeability, slope, extent of erosion, degree of wetness and climate conditions, particularly the rainfall, are used to determine land capability classification (LCC). A brief description of land from each capability class is presented below:

The land capability classification scheme developed by soil conservation service.

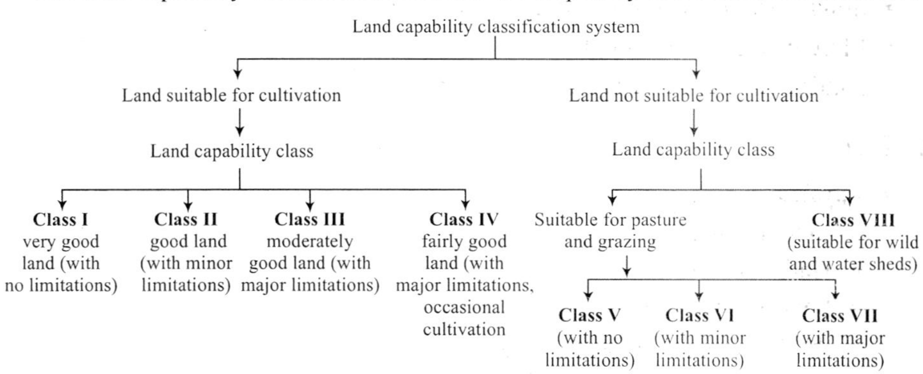

A. Lands suitable for cultivation

- **Class I (Light green):** Class I lands have very few or no limitations for growing common field crops. These are nearly level and the erosion hazard is low. The soil is deep, well drained, well supplemented with plant nutrients and responsive to application of fertilizer. Climate is favourable for growing a number of field crops. Availability of irrigation also removes the limitation by low rainfall in an arid region.
- **Class II (Yellow):** These lands have some limitations which reduce the choice of crops and require simple conservation practices. The limitations may result from gentle slope, slight susceptibility to erosion, less than ideal soil depth, wetness which can be corrected by drainage, slight to moderate salinity or sodicity that can be corrected easily, a slight climatic limitation etc. The limitations are very few and corrective measures are easy to apply.
- **Class III (Red):** These lands have more limitations than those in class II lands. The limitations of class III lands may result from one or more factors like moderately sloping land, moderate susceptibility to erosion, slow permeability, wetness or continuing water-logging even after drainage, shallow soil depth, low moisture holding capacity, moderate salinity/sodicity, and moderately limiting climatic conditions etc. Conservation practices are more difficult to apply.
- **Class IV (Blue):** Class IV lands have severe limitations that restrict choice of crops and require careful management. The restrictions in use of these lands are more than those for class III lands. The choice of cultivated crops is limited as a result of permanent features such as steep slope, severe susceptibility to erosion, severe effects of erosion, shallow soil depth, poor water holding capacity, excessive wetness with continuing hazards of water logging even after draining, severe salinity or sodicity, moderately adverse climate etc. Very careful management is required if crops are grown at all. Conservation practices are more difficult to apply and maintain.

B. Lands not suitable for cultivation

- **Class V (Dark green):** Although class V lands are nearly level and are not susceptible to much erosion, cultivation is not feasible for one or more limitations such as stoniness, wetness or severe climate. Class V lands are not suitable for raising cultivated crops but are suitable for perennial vegetation (Grazing and forestry). It is practical to apply pasture improvement practices such as seeding, application of lime or gypsum, application of fertilizer, rain water harvesting, provision of drainage etc.
- **Class VI (Orange):** Class VI lands have very severe limitations that make them generally unsuitable for cultivation. They have continuing limitations which cannot be corrected such as a steep slope, very severe erosion hazard, very serious effect of past erosion, stoniness, shallow rooting-zone, excessive wetness, low moisture holding capacity, salinity or sodicity and severe climate. Soils in this class are subject to moderate limitations under grazing or forestry use.

- **Class VII (Brown):** Soils in class VII lands have very severe limitations that make them completely unsuitable for cultivation and restrict their use largely to grazing and afforestation. The physical condition of these lands are such that it is practicable to adopt pasture improvement and water harvesting practices.
- **Class VIII (Purple):** The soils of class VIII lands are such that they can be used for commercial plant production and restrict their use to recreation, wild life food and cover, watershed protection or aesthetic application. The limitations which cannot be corrected may result from one or more of factors like severe erosion, severe climate, permanent wetness, stoniness, severe salinity or sodicity. It is, however, obvious that lands with higher LCC ratings (poorer lands) can be improved and converted to lower LCC (better lands) by adopting suitable reclamation practices.

33. SOIL NUTRIENT MINERALISATION

Soil Nitrogen: Nitrogen is a key element in plant growth. Despite large quantities of nitrogen (in the form of N_2 gas) in the atmosphere, the nitrogen cannot be utilized by majority of plants as such, therefore it is changed to a form suitable for use by plants. It is absorbed by cationic form as ammonium (NH_4^+) or anionic form as nitrate (NO_3^-). Only small part of soil nitrogen occurs in these forms whereas large amount remains in organic matter as amine. On decomposition of organic matter, by a complex biochemical processes, the organic nitrogen changes to ammonium salts, the process is called ammonification. Then ammonium nitrogen is oxidized to nitrate and nitrate forms by the oxidative process of nitrification. These transformations are carried out by soil microflora. The bacteria, *Nitrosomonas* and *Nitrococcus* change ammonium in to nitrite and nitrobacter changes nitrite to nitrate.

Nitrogen fixation: The process in which the atmospheric nitrogen (N_2) is converted to a form used by plants and micro organisms, is called nitrogen fixation. Nitrogen can possess a variety of oxidation such as :

$$\underset{\text{Nitrate}}{NO_3^-} \longrightarrow \underset{\text{Nitrite}}{NO_2^-} \longrightarrow \underset{\text{Hyponitrite}}{N_2O_2} \longrightarrow \underset{\text{Nitrogen gas}}{N_2} \longrightarrow \underset{\text{Hydroxylamine}}{NH_2OH} \longrightarrow \underset{\text{Ammonia}}{NH_3}$$

Thus in nature, nitrogen may exist in either highly oxidized form (NO_3^-) or highly reduced form (NH_3).

A. Non-biological fixation: During lightning storm and electrical discharge, extreme temperature and pressure are created which oxidize nitrogen of atmosphere to nitric oxide or nitrate.

$$N_2 + O_2 \xrightarrow{2O_2} 2NO \longrightarrow 2NO_3^-$$

The oxides of nitrogen hydrated by water vapour and is carried over to earth by rain. It is estimated that more than 200000 tons of nitrogen is fixed, every year, by this process.

During the first World War, German Scientist Haber succeeded to fix (convert) atmospheric nitrogen to ammonia at high temperature and pressure.

$$N_2 + 3H_2 \xrightarrow[200\ atm]{450°C} 2NH_3 \longrightarrow H + 24\ K\ calorie$$

This process is now–a–days used by chemical industries in production of commercial fertilizers.

B. Biological Fixation: Major amount of nitrogen is fixed by biological sources. Among the biological sources, two groups of organisms are actively involved in nitrogen fixation.

(i) Nonsymbiotic organism: The free living microorganisms exist independently in soil, which fix nitrogen in their body in the form of organic molecules. They have short span of life, release the organic nitrogen when they die and on decomposition, the nitrogen becomes available to the plants in the form of ammonium, nitrate or nitrite. The non-symbiotic fixation of nitrogen is estimated over 25 kg per hectare with an average of 5-8 kg per hectare annually. The non-symbiotic organisms are divided into following groups.

- Anaerobic organism – *Clostridium species*
- Aerobic organisms – *Azotobacter spp.*
- Photosynthetic organism – *Rhodosprillium rubrum*
- Algae – *Myxophaceae*

The nonsymbiotic organisms obtain energy from decomposing organic matter of soil or from photosynthesis. Therefore, for non-symbiotic nitrogen fixation the soil must have appreciable amounts of decomposable organic matter. *Azotobacter* and Clostridium species are of economic importance which fix atmospheric nitrogen approximately 20-50 kg per hectare annually in organic matter rich soils. The photoautotrophic organisms such as algae, *Rhodosprillum* are actively involved in fixing nitrogen in soil. These organisms have enzyme system, the nitrogenase which convert nitrogen easily into organic form as amine, NH_2 (in amino acids).

(ii) Symbiotic organisms: In many species of leguminous plants bacteria (Rhizobium) and *Actinomycetes* cause the formation of root nodules. These organisms live symbiotically by taking their food and energy from host plant and fix atmospheric nitrogen in return. Symbiotic nitrogen fixation of nitrogen by legume bacteria, *Rhizobium* can add 50- 280 kg nitrogen per hectare per year. Leguminous plants benefit soil markedly from their symbiotic relationship with rhizobium and consequently the soils in which they are grown also get more benefit, when these crops are turned into soil.

Mineralisation of nitrogen: The major source of soil nitrogen is organic material (by added organic matter or symbiotically and nonsymbiotically fixed nitrogen). When soil organic matter is decomposed, nitrogen is released as the usable nutrient ions, ammonium and nitrate. This microbial conversion process is called mineralization.

This process is divided in two stages, in first, organic nitrogen is changed to ammonium salts called ammonification and in second stage, ammonium compound is changed to nitrite or nitrate salts, called nitrification.

$$NH_4^+ + 3O_2 \xrightarrow[\text{Nitrification}]{\text{Enzymes}} 2NO_2^- + 2H_2O + 4H^- + \text{Energy}$$

$$2NO_2^- + O_2 \xrightarrow[\text{Nitrification (Nitrobacter)}]{\text{Nitrosomonas, Nitrosococcus}} 2NO_3^- + \text{Energy}$$

The bacteria which are responsible for these changes are called nitrifying bacteria or nitrifiers. The first change of ammonium ion to nitrite is carried out by *Nitrosomonas* and *Nitrosococcus*, while the bacteria involved in the oxidation of nitrite are called Nitrobacter.

Denitrification: The loss of nitrogen from soil in gaseous form is also known as denitrification. This occurs as a result of biological processes in which microorganism participate and as a result of chemical process that take place either in association with the biological transformation or independently of them. The nitrogen present in the organic residue is removed as NH_3 by oxidation of nitrogenous organic compounds specially proteins to carbon dioxide and water.

Aerobic microorganisms when do not get oxygen as per their demand, specially under waterlogged conditions, the nitrate nitrogen is subjected to a microbial attack resulting the release of molecular nitrogen.

$$NO_3^- \xrightarrow{\text{Microbial/Enzymatic}} NO_2^- \text{ or } N_2O \text{ or } N_2\uparrow$$

$$NO_2^- \xrightarrow[\text{Anaerobic}]{\text{denitrification}} N_2O \text{ or } N_2\uparrow$$

In acid soils if ammonium nitrate is formed in appreciable amount there is a strong possibility of loss of gaseous nitrogen through its decomposition. On the other hand, in alkaline soil there is more possibility of reaction with ammonium ions and loss of nitrogen is due to liberation of ammonia.

B. Soil Phosphorus: Plants absorb most of their phosphorus as the primary orthophosphate, H_2PO_4 and smaller amounts as secondary orthophosphate ion, HPO_4^{--}.

The kind of phosphorus present in solution depend upon the pH of the soil solution. In acidic (pH 5.5) conditions, the $H_2PO_4^-$ ion is dominant, while with increasing pH, HPO_4^- ions are increased and in distinctly alkaline soil (pH 7.4) PO_4^{---} ions are present in excess. Moreover at low soil pH (below 5.0), Fe, Al, Mn, Ti-oxides become soluble (Colloidal), which attract phosphate and make it insoluble as Fe, Al, Mn and Ti-phosphates. In this way phosphate is fixed in soil.

$$Al^{+++} + H_2PO_4^- + 2H_2O \longrightarrow 2H^+ + Al\,(OH)_2\,(H_2PO_4) \text{ insoluble}$$

On the other hand, pH values above 7.5 and abundant lime form complex as calcium phosphates which result to increase the insolubility of phosphates e.g. Oxyapatite, $[3Ca_3(PO_4)_2\ CaO]$ and fluorapatite $[3Ca_3(PO_4)_2\ CaF_2]$. At pH above 8.0 or 8.5, the solubility of phosphorus decreases extensively. The native phosphorus occurs in the soil by disintegration of rocks which are fluorapatite, hydroxyapatite, chlorapatite or as iron phosphate and aluminium phosphate.

The water soluble orthophosphate ions in an acid medium favours the formation of monophosphate ion, $H_2PO_4^-$ in neutral to mild alkali media, the formation of di-phosphate ion HPO_4^- and in alkaline medium, the formation of triphosphate PO_4^{---} occurs.

On decomposition of organic matter CO_2 gas is produced, which dissolves in water to produce carbonic acid, H_2CO_3. This weak acid solubilises phosphorus to a large extent. The humic acid forms complex with phosphorus as phosphohumic complex. A close relationship between sesquioxide and phosphate, R_2O_3:P_2O_5 also exists in soil.

The organic phosphorus is decomposed or mineralized by the activity of microorganisms. The various enzymatic actions convert the unavailable organic phosphorus into available form. Inorganic phosphorus added to the soils is converted to organic compounds by soil microorganisms. This balance proceeds in a ratio of mineral phosphates: organic phosphate, if the ratio is 200 or less, mineralization predominates while if ratio is more than 300, immobilization is favoured.

34. INTERATED PLANT NUTRIENT MANAGEMENT (IPNM)

Integrated Plant nutrient management is the combined application of chemical fertiizers along with organic manures, green manures, biofertilizers and other organic recyclable materials for crop production. The basic concept underlying is as below:

- IPNS is the maintenance or adjustment of soil fertility and plant nutrient supply to an optimum level for sustaining the desired crop productivity through optimization of the benefit from all possible sources of plant nutrients in a integrated manner.
- IPNS is ecologically, socially and economically viable and environmentally healthy technology, which can be adopted by the farming community to get higher productivity while maintaining the soil fertility.
- IPNS improves the physical, chemical and biological conditions of the soil.
- IPNS encourages the use of in-house organic waste; thus it saves on the cost of fertilizer for crop production. The proper management of water also helps in keeping clean environment.
- IPNS improves the water holding capacity of the soil under dryland conditions, thus risk of crop failure is minimized.
- IPNS reduces/minimizes infestation of insects/pest and diseases in the crops.
- Increases organic matter content and physical conditions of soil.
- Increases efficiency of added chemical fertilizers by its temporary immobilization which reduces leaching of plant nutrients.
- Increases availability of plant nutrients for a longer period.
- Improves the quality of produce i.e. protein, oil percentage etc.
- Reduces hazardous effect on soil health.
- It increases the beneficial microorganism in the soil.
- It reduces the pollution effect.
- It has synergistic effect on improvement in soil health and overall productivity of the system.
- It minimizes the nutrients fixation in the soil.

Some important conversion factors:

- $N \times 1.286 = NH_4$
- $N \times 4.43 = NO_3$
- $NO_3 \times 0.226 = N$
- Organic carbon × 1.724 = Organic matter
- $P \times 2.29 = P_2O_5$
- $P_2O_5 \times 0.437 = P$
- $K \times 1.20 = K_2O$
- $K_2O \times 0.83 = K$
- $Ca \times 1.40 = CaO$
- $CaO \times 0.7147 = Ca$
- $Ca \times 2.497 = CaCO_3$
- $CaCO_3 \times 0.4004 = Ca$
- $Ca \times 1.85 = Ca(OH)_2$
- $S \times 3.00 = SO_4$
- $SO_4 \times 0.3333 = S$
- $Mg \times 1.658 = MgO$
- $MgO \times 0.604 = Mg$
- $Mg \times 3.468 = MgCO_3$
- $MgCO_3 \times 0.2884 = Mg$

General sufficiency or optimum range of macro and micro nutrient elements in plants

Nutrients	Sufficiency or optimum range
Macro nutrients (%)	
N	2.0-5.0
P	0.2-.05
K	1.0-5.0
Ca	0.1-1.0
Mg	0.1-0.4
S	0.1-0.3
Micronutrients (μg/g)	
Zn	20-100
Fe	50-250
Mn	20-300
Cu	5-20
B	10-100
Mo	0.1-0.5
Cl	2000-20000

Rating chart for soil test values of some nutrients

Content	Low	Medium	High
Organic carbon	<0.5	0.5 – 0.75	>0.75
Available N (kg/ha)	<280	280 – 560	>560
Available P (kg/ha)	<10	10 – 25	>25
Available K (kg/ha)	<120	120 – 280	>280
Sulphur (kg/ha)	<18	18 – 36	>36
Zn (ppm)	<0.6	0.6 – 1.2	>1.2
Fe (ppm)	<4.0	4.0 – 8.0	>8.0
Cu (ppm)	<0.2	0.2 – 0.4	>0.4
Mn (ppm)	<2.0	2.0 – 4.0	>4.0

35. ORGANIC FARMING

Organic farming is the method of crop and livestock production that involves much more than choosing not to use pesticides, fertilizers, genetically modified organisms, antibiotics and growth hormones. In other words, it is referred to as low input farming and uses natural sources of nutrients such as compost, crop residues and manure and natural methods of crop and weed control, instead of using synthetic or inorganic agrichemicals. Organic farming seems to be more appropriate as it considered the important aspects sustainability. Organic farming avoids the use of chemicals and protects the environment from pollution. Organic farming promotes eco-friendly agricultural practices without making use of synthetic inputs and may only relies upon the use of organic wastes to raise crops. Organic farming seems to be more appropriate as it considered the important aspects live sustainable natural resources and environment. It is a production system, which favours maximum use of organic materials like crop residues, FYM, compost, green manure, oil cakes, bio-fertilizers, bio gas slurry etc. to improve soil health. Organic farming is a holistic way of farming: besides production of goods of high quality, an important aim is the conservation of the natural resources fertile soil, clean water and rice biodiversity. Organic farming should sustain and enhance the health of soil, plant, animal and planet as one and individual.

Organic farming endorse the concepts that the soil, plant, animals and human beings are linked. The benefits of organic farming rely on crop rotation, crop residues, animal manures, legumes, green manure, off farm organic wastes and biofertilizers, mechanical cultivation, mineral bearing rocks. The agricultural products can be produced organically, including grains, meat, dairy, eggs, fibers such as cotton, jute, flower etc. Organic farming can actually reduce its production cost by 30% as compared to the cost of conventional farming. This eliminate the use of expensive synthetic fertilizers and pesticides, minimizing soil corrosion by up to 55% and increase crop field. Besides this, the use of organic farming microbial biomass can be increased up to 20–30% and microbial activity can increased 30–100%. The physical properties of soil such as soil structure, bulk density, porosity, water holding capacity, soil aeration, soil temperature etc. are positively improved by the adoption of organic farming.

Objectives of organic farming

There are following objectives of organic farming:

1. Ensure the right soil cultivation at the right time.
2. Keep and build good soil structure and fertility.
3. Increase genetic diversity.
4. Promote more usage of natural or bio pesticides.
5. Control insects, pests, diseases and weed infestation.

Principles of organic farming

The important principles of organic farming are as:

1. **Principles of health:** Organic agriculture must contribute to the health and well being of soil, plants, animals, human and the earth.
2. **Principles of ecological balance:** Organic farming should be based on living ecological balance and cycles in nature.
3. **Principles of fairness:** Organic farming provides good quality of life and helps in reducing soil infertility.
4. **Principles of core:** Organic farming should be managed in a precautionary and responsible manner to protect the health and well being of current and future generations and the environmental.

Types of organic farming

There are different types of organic farming:

1. **Pure organic farming:** It involves the use of organic manures and bio pesticides with complete avoidance of inorganic chemicals and pesticides.
2. **Integrated organic farming:** It involves integrated nutrients management and integrated pest management. It is the type of farming in which development of crops from natural resources having the complete nutritive value and manage to prevent the crop or plants from the pests.

In organic farming following techniques used:

1. **Crop rotation:** It is the technique to grow various kind of crops in the same area, according to the different seasons, in a sequential manner.
2. **Green manure:** It refers to the dying plants that are uprooted and turned into the soil to make them act as a nutrient for the soil to increase its quality.
3. **Compost:** Highly rich in nutrients, it is a recycled organic matter used as a fertilizer in the agricultural farms.
4. **Biological pest control:** With this method, use of living organisms to control pests with or without use of chemicals.

Methods of organic farming

1. **Soil management:** After cultivation of crops, the soil loses its nutrients and its quality depletes. Organic agriculture initiates the use of natural ways to increase the fertility of the soil. Hence it focuses on the use of bacteria that is present in animal waste which helps in making the soil nutrient more productive and fertile.
2. **Weed management:** Weeds in the unwanted plant that grows in agricultural fields. Organic agriculture focuses on lowering weed rather than removing it completely. The two most widely used weed management techniques are–
    - **Mulching:** A process where we use plastic films or plant residue on the surface of soil to minimized the growth of weeds.
    - **Moving or cutting:** Removal of weeds top growth.
3. **Crop diversity:** Monoculture in the practice used in the agricultural fields where we harvest and cultivate only one type of crop in a particular location. Recently, polyculture has come in existence, where we harvest and cultivate different kinds of crops to meet the increasing crop demand and produce the required soil microorganisms.
4. **Controlling other organisms:** There are both useful and harmful organisms in the agricultural farm which affect the field. So we need to control the growth of such organisms to protect the soil and the crops. For controlling of these organisms the use of herbicides and pesticides that contain fewer chemicals or are natural.

Importance of organic farming

- The environmental benefits because natural habitat sources are less threatened.
- It provides healthier food for people.
- The soil is in better conditions because of the manure or compost used.

Advantages of organic farming

There are several advantages of organic farming are:

1. Organic farming helps to prevent environmental degradation and can be used to regenerate degraded areas.
2. Organic manures produce an optimal condition in the soil for more yields and good quality of crop produce.
3. They minimized the need for purchased inputs.
4. Poison-free produce can be obtained.
5. They improve the soil chemical properties such as supply and retention of soil nutrients and promote favorable chemical reactions.
6. Organically grown plants are more resistant to diseases and insects and hence only a few chemical sprays or other protective treatment are required.
7. It generate the employment opportunities.

8. It increases the total benefit of the farm and increases the economic status of the farmers.
9. It increases the beneficial soil microorganisms in the soil.
10. It regulate the pH and electrical conductivity of the soils.
11. It act as a soil amendment to improve soil physical, chemical and biological properties of soil.

Organic farming has clear benefits in reducing environmental pollution, increase crop productivity and income, improvement in soil quality, high nutritive value, increases in population of beneficial insect and employment opportunities etc.

36. BIOFERTILIZERS

Biofertilizers are quite inexpensive and their use is very simple so that average farmers can easily apply. The cost of bacterial and BGA biofertilizers is Rs. 10 which is enough for one acre of cropped area. Biofertilizers can effectively reduce the requirement of crops to nitrogenous and phosphatic fertilizers. The biofertilizers are becoming very popular due to increased emphasis on organic farming and also due to high cost of synthetic or inorganic fertilizers.

Biofertilizers are indeed distinct from the chemical fertilizers in several ways including their mode of nutrient apply. It should be clearly understood that they themselves do not provide any nutrients to the crop. It is only their functions and interactions in the rhizosphere of the plants that increases the nutrient supply for uptake by the crop.

"Biofertilizers are living microorganisms capable of colonizing the root surface and providing benefits to the crops while utilizing the carbon supply made available to them in the form of root exudates."

Unlike cereals, pulses thrive maximum benefit from these microorganisms. Pulses form nodules in the roots. Each nodule in root is a small scale factory for the transformation of atmospheric nitrogen to the plant available form. This transformation is carried out by a specific microorganism commonly called as rhizobium.

The term biofertilizers is therefore commonly used for the commercial preparation of all these beneficial microorganisms. These preparation are commercially available in different forms. Some are solid while others are liquid based preparations. Liquid based biofertilizers have longer shelf life and easier to apply. Solid based preparations could be a powdered or granulated form. Normally, charcoal based or lignite based preparations are also available. Storage and shelflife of liquid preparations is however better than solid carrier based preparations.

Important biofertilizers are described below:

(a) Bacterial biofertilizers:

- ***Rhizobium*** help in establishing efficient symbiotic association with pulse and leguminous fodder crops and thus can fix 50 to 100 kg N/ha and also leave sizeable amount of nitrogen in the soil to meet a part of requirement of succeeding crop in rotation. The harmful effects of acidity or alkalinity are minimized due to rhizobium inoculation. There are several species of rhizobium specific for different pulses. Therefore, it should be understood that single rhizobium inoculant should not be used in all pulses. It means that rhizobium for chickpea would not work satisfactorily with spring or summer pulses like mungbean and *urdbean*. Rhizobium cultures for following crops can be made available to farmers like lentil, berseem, groundnut, pigeonpea, green gram, pea, *dhaincha*, gram, lucerne, blackgram, soybean, cowpea etc.

Rhizobium species and strains of different group

Group	Species of rhizobium	Name of crop	Rhizobium strain
Pea group	*Rhizobium leguminosorum*	Pea, lentil, lathyrus, chickpea, *kulthi*	P-12, HUPRP, 10-76, M-1-4 LC-1, LC-4, D K-3,4,7,TAL634 L-1-77, L-21-83 Khesari K-1, K-3, K-5, DK 5,6,7 Chana K 101, G 10-18 RC 149, H-45 DK -4, G 72 FIC 76 Kulthi VPR-1, AL 3, HVD 6, MKPAG-1
Bean group	*Rhizobium phaseoli*	*Arhar*	BC 1021, A-11, CC-11, IC 3100, IHP -195
		Mung	M-10, M-11, MO 5, KM -1, GMG 10, VS-1, S-24
		Urd	M-11, S 24, KU 1, BKM, BS-1
		Lobia	CC 3, DC 28, DC 6, K3, P 109 UASIVS
		Guar	G-3
Lucern group	*Rhizobium meliloti*	*Lucern*	SU 216
Soybean group	*Rhizobium japonicum*	*Soyabean*	Stain 1028
Lupin group	*Rhizobium lupini*	*Lupin*	
Clover group	*Rhizobium trifolii*	*Berseem*	21-19 C

- **Azotobacter** are recommended for seed inoculation of cotton, wheat, maize, sorghum, mustard and vegetable crops. Applications of these inoculants provide a 15 – 20 kg of biologically fixed nitrogen to the crops. Azotobacter inoculants also act as growth promoting substances which help in better seed germination and root proliferation to the crop and also protect the crop from fungal diseases.

Thus azotobacter plays nutritional, stimulatory and therapeutic role for the plant growth. Seed or seedling bacterization with Azotobacter cultures is a very simple process. One packet of culture is sufficient to inoculate seed material to sow an *acre* of land. In case of seedlings, the roots are to be dipped in inoculant slurry.

- **Azospirillum** are found beneficial and are recommended for sorghum, millets, barley, oats and wheat. Azospirillum inoculant could lead to yield improvement equivalent to those obtained with the application of 30 kg of fertilizer nitrogen per hectare. One packet of the inoculant consisting of 200 kg is enough for treating seeds sufficient for sowing one *acre* of land.
- **Phosphate solubilizing biofertilizers:** In addition to the rhizobium, there are other beneficial microorganisms which increase the supply of nutrients other than nitrogen such as phosphorus. These microorganisms are collectively called as Phosphate Solubilizing Bacteria (PSB). The introduction of the most efficient phosphate solubilizing microorganism in the rhizosphere of crop and soil increases the availability of phosphorus from insoluble sources of phosphate and also enhances the efficiency of phosphatic fertilizers. The phosphate, which gets fixed into soil, is also made available more effectively to *crops* by these microorganisms. A large number of microorganisms such as *Pseudomonas striata, Bacillus polymyxa* and *Aspergillus awamori* have been found efficient in solubilizing low grade mussorie rock phosphate and also soil fixed phosphorus. About 40% of superphosphate could be saved by the application of rock phosphate and phospho-microorganism. Phospho-microorganisms could be used for all crops. This helps in greater P uptake and increases the crop growth and yields.
- **Plant Growth Promoting Microorganisms (PGPR):** Another group of microorganisms that has shown potential to increase crop yields through production of phyto-hormones and other biomolecules involved in the control of soil borne diseases are collectively called as plant growth promoting Microorganisms (PGPR). Use of PGPR in pulses increase growth alongwith rhizobium and PSB cultures. Combined inoculation of these beneficial microorganisms has been found to increase the yield by 10-20 % compared to uninoculated controls.

  In addition to these microorganisms, certain fungi such as *Trichoderma spp* and VAM fungi are also being produced in large amounts for use as biofertilizers. These fungi are known to suppress soil borne diseases and also increase nutrient availabilty.

Types of microbial inoculants

A. Nitrogen fixers

(i) Symbiotic : Rhizobium inoculants for legumes

(ii) Non symbiotic: For cereals, millets, vegetables

1. **Bacteria**

(a) **Aerobic:** *Azotobcter, Azomonas, Azospirillum, Mycobacterium*

(b) **Anaerobic:** *Clostridium, Chlorebium, Chromatium*

(c) **Facultative anaerobes:** *Bacillus, Enterobacter, Eischerichia, Klebsiella, Rhodospirillum*

2. **Blue green algae:** *Anabaena, Anabaenopsis, Nostoc, Tolypothrix*

(b) **Phosphate solubilizing microorganisms:** *Pseudomonas, Bacillus* (bacteria), *Aspergillus, Penicillium* (Fungi)

(c) **Cellulolytic microorganisms:** *Chaetomium, Trichoderma* (Fungi), *Cellulomonas, Clastridium* (Bacteria), *Nocardia, Streptomyces* (Actinomycetes)

(d) **Sulphur dissolving bacteria**

(e) **Azolla:** *Anabaena azollae, A. pinnata*

(f) **Lignolytic microorganisms:** *Clavaris, Cephalosporium, Humicola* (Fungi)

(i) **Recognised species of Azotobactor are:** *A. chroococcum, A. vinelandii, A. beijerinckii, A. paspalli, A. macrocylogenea, A. insignis, A. agilis*

(ii) **Recognised species of rhizobium are:** *R. leguminosarum* (pea group), *R. phaseoli* (bean group), *R. trifolii* (Clover group), *R. mililoti* (alfalfa group), *R lupin* (lupine group), *R. japonicum* (soybean group), *R. spp.* (cowpea group).

B. **Compost inoculants:** The recycling or utilization of farm wastes is essential for maintainance of soil health and replenishment of plant nutrients. One of the best methods of recycling of organic materials is through composting. The bulk of the component of organic materials is made up of cellulose followed by hemicellulose and lignin. The cellulolytic fungal cultures such as *Trichurus spinalis, Aspergillus niger, Paecilomyces fusisporus* and *Trichoderma viride* have been identified as promising ones. The inoculation of organic matter with any of fungi at 0.5 kg/tons resulted in hastening of the composting process. The culture should be applied in the beginning with water while filling the pit or making windows. As soon as a carbon nitrogen ratio of 20:1 or less than this is attained, compost is ready for use. The properties of the composted material are similar to farm yard manure. Besides rapid composting, the quality of product is also improved. It have been observed that the rice straw which has a nitrogen content of 0.4 % can be converted within 6-9 weeks into organic material containing about 1.6 per cent nitrogen. Due to inoculant, humus and available P_2O_5 content are also increased.

Organic material can be composted by different methods but the most convenient, inexpensive and simple method of composting is heaps or in pits. In case of pit method, the depth of the pit should be about 1 meter and the length and the breadth of the pit can be varied depending on the availability of organic materials to be composted. One cubic meter pit can accommodate about 100 to 125 kg straw. Being poor in nitrogen (0.5 % nitrogen) the carbon – nitrogen ratio of cereal residues is quite wide and for rapid composting the C:N ratio should be brought down to 60:1 either by blending with organic materials comparatively high in nitrogen such as leguminous residues, grass clippings, green leaves and weeds or with chemical nitrogen. Nitrogen enriched phospho compost can also be prepared from crop residues amended with dung, soil and FYM (8:1:1.05) along with rock phosphate, pyrite (10%) and nitrogen (1%). The moisture on an average should be maintained at 60-75%. The composting mass should be kept neither dry nor too wet. Turning of material atleast 2 to 3 times during the composting period at fortnightly or monthly intervals can provide aeration.

C. **Vesicular Arbuscular Mycorrhizae (VAM):** VAM is being produced in sterilized sand and soil mixture with suitable host. It contains large number of fungal spores and infected root pieces. VAM inoculant forms association virtually with all the land plants, field crops and trees. It has following advantages:

- VAM increases the plant growth and health by improving the availability of soil phosphorus and trace elements such as zinc, copper and iron.
- VAM enhances the uptake of applied fertilizer nitrogen, thereby fertilizer use efficiency of the crops is increased.
- It also improves the biological nitrogen fixation of legumes by improving nodule formation.
- VAM fungi are also good contributor to soil stability by improving plant cover and root proliferation due to increased nutrient uptake.
- It also improves the crop nutritional status during period of soil water deficiency thereby improves the drought resistance of plants.
- Reduces the symptoms incited by disease producing organisms (fungal and virus also).
- It also enhances population of rhizospheric microorganism (*Azospirillum, Azotobacter*, *Rhizobium* and PSM).

VAM should be applied in close proximity of the seed. It should be applied first before sowing of seed especially with the transplanted crops. Generally 2-4 kg of VAM fungi inoculant is recommended for one acre. This dose varies with the crops. For transplanted crops such as chilli, tomato, brinjal, rice etc., it should be spread over the nursery bed first then transplanting should be done. Similar techniques can also be used for other transplanted seedlings and trees also.

D. **Blue Green Algae (BGA):** Blue green algae also called cynobacteria are photosynthetic lower plants and synthesize energy rich compounds from atmospheric carbon dioxide and solar energy, hence are self supporting organisms. A large number of BGA strains such as *Anabaena* variable, *Nostoc maseorum, Aulasira fertilissima* and *Tolypothrix tenuis* are known to fix nitrogen and part of this fixed nitrogen is released during the growth phase and rest after death of autolysis. Apart from nitrogen fixation these organisms contribute to the productivity and fertility of soil through various ways and leads to increased crop production. These are:

- The process of nitrogen fixation contributes 25-30 kg fixed nitrogen/ hectare/season.

- Solubilization of inorganic insoluble phosphorus into soluble phosphorus which is available to the crop plants.
- Liberation of plant growth promoting substances like vitamin (Vitamin B 12, folic acid, nicotinic acid and pantothenic acid), auxins (indole acetic acid and 3-methyl indole), amino acids and sugars.
- BGA excrete complex organic carbon compounds like polysaccharides; these compounds bind the soil particles and improve soil aggregation.
- Improved soils structure results in improved infilteration rate, soil permeability and water holding capacity.
- In saline, saline-alkali and alkaline (sodic) conditions it brings the soil pH towards neutrality, reduces sodium content in soil and thus soils are reclaimed.
- Under acidic situations, pH has the tendency to rise towards neutrality.

With the use of BGA bofertilizer in rice crop beneficial effects observed through out the country are as follows:

- In areas where chemical N – fertilizer is not used for various reasons, algal inoculation enhances the crop yield by 5-30% in different agroclimatic regions.
- Even at the recommended levels of chemical nitrogen fertilizers being used, application of BGA biofertilizers results in increased crop yield by 10%.
- On comparing, the yield of rice with full recommended dose of inorganic fertilizers vis-à-vis at reduced levels, by application of BGA biofertilizers, a net saving of 20-30 kg N/ha is achieved without losing any yield.

Under field conditions, BGA can be applied in the following way:

- Broadcast the dry algal inoculum over the standing water in the rice field at a rate of 10 kg/ha. In clay soils water remains turgid for 2-3 days, hence BGA inoculum should be applied after 2-3 days of transplantation of rice seedlings. In loamy soils application should be done immediately after transplantation is completed. Application of excess BGA inoculum is not harmful and will accelerate the first growth of BGA in the fields within 2-3 weeks.
- Where chemical fertilizer should be used at high doses (100 kg N or more), the use of chemical fertilizers should be reduced by 25-30 kg N/ha and BGA inoculum should be used to get the same yield as at the full dose of chemical fertilizers.
- Recommended pest control measures and other management practices do not normally interfere with the establishment and activity of the BGA in the field.
- Use of BGA biofertilizers is mainly recommended for rice.
- Ensure that there is enough water atleast one week after BGA application.

Method of application: Inoculation can be done either of seeds of crops or seedlings before sowing. For inoculating seed for one *acre* of land, prepare 10% solution of jaggery (Gur) or sugar in water of quantity sufficient to moisten seeds of one *acre*. Boil the solution and cool it completely. Open the packet of the inoculant and mix these contents in the cooled solution thoroughly. Heap the seed on a clean surface preferably on polythene sheet and pour the inoculant slurry. Mix the slurry with seed evenly. Dry the inoculated seed in shade and sow immediately.

Now-a-days, Carboxy Methyl Cellulose is also being used as adhesive in place of jaggery or gum of Acacia. A thin layer of adhesive is uniformly spread over the seed prior to coating with biofertilizers powder. The packet of appropriate biofertilizers is opened and its powder is sprinkled over the seeds. Seeds are mixed with the powder slowly and gently to ensure uniform coating with the biofertilizers. These treated seeds are dried in the shade under a tree or suitable place away from the direct hot sun. Dried seeds are utilized immediately (before 6-8 hrs) for sowing. It has been observed that prolong storage of treated seeds for more than 24 h reduces the efficiency of the microorganisms in forming nodules and its activity. In case of PSB and PGPR cultures are also used alongwith the rhizobium culture, it is recommended that the commercial preparations are mixed in equal proportion prior to the coating on the seeds.

Precautions while using biofertilizers:

- The biofertilizers are living microorganisms therefore it should be ensured that the commercial preparations contain high population of living and efficient strains of rhizobium, PSB and PGPR microorganisms.
- A quality product of biofertilizers contains about 10^8 living cells per g of the carrier.
- The population of contaminated microorganisms should be less than 1000 per g carrier.
- Use the inoculant before expiry date, whenever applicable eg. Bacterial inoculants.
- Protect inoculant packet and the inoculated seed from direct sunlight and store in a cool place.
- Protect inoculated seed from direct contact with chemical fertilizers and pesticides.
- Sow the inoculated seed immediately (within 6-8 hrs).
- Always use appropriate inoculant for the crop.
- If fungicide or insecticide are used for seed treatments, the biofertilizers should be coated in the manner of fungicide than insecticide than rhizobium (FIR).

Advantages of biofertilizers:

- Low cost inputs
- Provides multiple benefits especially in increasing soil health
- Increases availability of plant nutrient in soil
- Most environmental friendly (ecofriendly)

Biological Nitrogen Fixation (BNF)

Nitrogen is an essential plant nutrient. The nitrogen is most commonly deficient in soils, contributing to reduced agricultural yields throughout the world. Nitrogen can be supplied to crop either through chemical fertilizers or biological nitrogen fixation. BNF is a process which is becoming more important not only reducing energy costs, but also in breaking more sustainable agricultural systems.

Average amount of nitrogen fixed by various legumes are described below:

Legume crops	Nitrogen fixation (kg N/ha/year)
Groundnut	100
Chickpea	105
Soybean	120
Stylosanthese hamata	135
Mungbean/Green gram	205
Cowpea	213
Pigeonpea	220
Lucern	250
Leucaena	350

For poor farmers, BNF is still a viable and cost effective. It is alternative or complementary solution to industrially manufactured N fertilizers. In addition to their great economic interest and indirect impact on human health, most BNF technologies have the potential to generate global environmental benefits by reducing green house gas emissions and water pollution and promoting more sustainable use of agricultural land.

Strategies to enhance BNF in agricultural system

These are several methods available for enhancement of nitrogen fixation:

1. Host plant management
2. Selection of effective strains able to fix more nitrogen
3. Use of different agronomic methods that improve soil conditions for plant and microbial symbionts
4. Inoculum production and inoculation methods

1. Host plant management:

(i) **Plant selection:** The amount of nitrogen fixed by legumes varies widely with host genotype, rhizobium efficiency, soil and climatic conditions. The effectiveness of various legume species and their micro-symbionts has been well established through various studies. The nitrogen fixing potential of a number of different legume species and their microsymbionts has shown that when effective rhizobial populations are present either naturally or from inoculation, and there are no other major yield limiting factors, plant selection is a potential method to enhance BNF. The high values for nitrogen fixation were observed on adapted genotypes and breeding lines when the environmental conditions were favourable.

(ii) Plant improvement:

1. **Genomics and new opportunities for BNF technologies**: Genomics is revolutionizing research and commerce in life sciences and offers the development of a new paradigm in the application of BNF technologies. Despite the whole genome sequencing of rice and Arabidopsis, legume species such as *Lotus japonicus* and *Medicago truncatula* have recently been chosen as model systems for genomics and functional genomics, particularly to accelerate the research on symbiotic nitrogen fixation. The scientific emphasis of this research will generate a global understanding of the expression, regulation, dynamics and evolution of the symbiotic properties, which will certainly open new opportunities for application and result in high socio-economic impacts.

   Although the direct molecular modification of a host plant or microsymbiont has to result in the improvement of N fixation at the field level, several approaches offer promise and some of those are:

   - **Host transformation to modify host range**: In a recent study transgenic lotus plants transformed with the soybean lectin gene became susceptible to infection by *Bradyrhizobium japonicum*.
   - **Host modification to synthesize opines:** Because rhizobium strains vary in their ability to use opines, genetic engineering of legumes or other plants for opine synthesis may result in the enhancement growth of rhizosphere organism with the ability to utilize this substrate.
   - **Genetic transformation of plants for enhancement malate dehydrogenase (MDH) synthesis in roots and nodules:** Malate is the primary plant carbon source used by bacteroid, and is also a factor in plant adaptation to P and Al stress. Alfalfa transformed with a MDH gene having high efficiency in malate synthesis, exuded more organic material into the rhizosphere and fixed more nitrogen than the wild type.
   - **Host mutants with improved characters:** Host mutants with improved characters such as disease and insect pest resistant, earlier and later flowering produced higher yield, higher protein content or less toxic compounds.

2. **Microsymbiont – Rhizobia:** Several important characteristics to be considered for the selection of the rhozobial symbionts are:

   (a) Nitrogen fixing ability: The rhizobia involved in nodulation can influence the percentage and amount of nitrogen fixed by the legume/rhizobium symbiosis. These are several methods available to quantify and estimate nitrogen fixation. Plant dry weight is usually well correlated with effectiveness in nitrogen fixation, when nitrogen is the only limiting growth factor. 15 N-based methods provide direct evidence for nitrogen fixation.

(b) **Competitive ability:** The proportion of the nodules formed on a particular host is influenced by the competitive ability of an inoculated rhizobium strain in comparison to indigenous strains, which may vary in their effectiveness.

(c) **Thermotolerance and survival ability**: Evidence for the involvement of some plasmids in imparting resistance to the adverse condition in some strains of rhizobium was documented. Survival of these strains in soil can be improved by identification of plasmids carrying genes for resistance to environmental stress and transferring them to the highly effective strains.

Characters desirable for the strains for use in commercial inoculants

(i) Ability to form nodules and fix nitrogen on the desirable host.
(ii) Ability to compete in nodule formation with ineffective to native population.
(iii) Ability to fix N across the range of environment.
(iv) Ability to form nodule and fix nitrogen in the presence of soil nitrates.
(v) Ability to grow well in artificial media, inoculants carrier and in soil.
(vi) Ability to persist in soil particularly for seasonal legumes.
(vii) Ability to migrate from the initial site of inoculation.
(viii) Ability to colonize in the soil in the absence of host legume.
(ix) Ability to tolerate environmental stress.
(x) Ability to fix nitrogen with wide host genotypes.
(xi) Genetic stability.
(xii) Compatibility with agro-chemicals
(xiii) Low mortality on inoculated seed.

3. **Microsymbiont-Mycorrhiza (VAM fungi):** VAM fungi are obligate symbiont associated with majority of the agricultural crops growing under broad ecological range. Many graminaceous plants and legumes are highly susceptible to VAM colonization. The transfer of nutrients mainly P from the soil to the cell to the root cortex is mediated by intercellular obligate endosymbionts of the genera *Gloma, Giyapsora, Acaulospora, Sclerocystis* and *Endogone* which possess the vesicles for storage of nutrients and arbusculer for funneling of these nutrients to the plant root system. The fungi produce phosphatases which allow the utilization of the organic phosphorus. VAM also has been reported to improve the uptake of micronutrients such as Zn, Cu, S, Fe and water. P and micro nutrients are the key elements for enhancing the BNF; by the way of enhancing the uptake and availability VAM symbiosis indirectly improves BNF in legume. Benefits of VAM fungi highly relate to their amount of infection in host plant. Cultivars also play an important role on VAM symbiosis in crops.

37. SOIL AND WATER POLLUTION

The soil environment plays an important role in the circulation of substances and maintenance of ecosystem. It performs functions such as soil and water purification and the production of food and wood. In recent years, the soil and water pollution has increased rapidly. The moist common soil contaminants are organic chlorides such as trichloroethylene.

A. Contamination of soil with pesticides: Pesticides can be classified into following three group.

1. Insecticides: Chemical responsible for controlling of insects is known as insecticide. Insecticide may be classified into chlorinated hydrocarbons, organophosphate, Carbonate and plant origin.

(i) Chlorinated hydrocarbons: This group of insecticides has low biodegradability. They persist in the soils and plants for long period and cause toxicity e.g., DDT, BHC, Lindance, Aldrin, Endrin etc.

(ii) Organophosphorus: The organophosphorus insecticides are Parathion, Malathion, Metasystox, Phosdrin etc generally biodegradable and thus, less likely to build up in soil and water. Systemic insecticides of this group are relatively much more toxic to human and other mammalians. Great care should be taken not to inhale and expose any part of body at the time of spraying.

(iii) Carbonate: Carbonate insecticides like carbofuran, carbaryl, oxamyl, seven etc are easily biodegradable and relatively less toxic.

(iv) Plant origin: Plant origin insecticides which are extracted from plants are known as plant insecticides *viz.*, nicotine, pyrethrum, nimbidin etc. Such insecticides unlike synthetic insecticides are safer to use.

2. Fungicide: Fungicides are chemicals used for controlling of diseases of crops. Some fungicides containing mercury, arsenic and copper are poisonous to man and animals. Organic fungicide such as dithiocarbamate (Zinam, Zineb, meneb, dithane, ferbam etc.) are less toxic.

3. Herbicides: The chemicals which kill or inhibit the growth of weeds are called herbicides or weedicide. Weedicides are generally less toxic than insecticides and fungicides. Some are quite toxic to water animals like fishes. Weedicides are generally biodegradable e.g., 2, 4-D, Trillate, Eptam etc.

B. Contamination with toxic inorganic compounds: Soils are also contaminated by a number of inorganic compounds. They are also toxic to human and animals. Arsenic, cadmium, mercury and chromium are extremely toxic, lead, fluorine and nickel are moderately, zinc, copper, boron and manganese are relatively less toxic.

Mazor sources of toxic inorganic compounds

Highly toxic

Arsenic: Pesticides, coal, petroleum etc.

Cadmium: Fertilizer impurities, paints, batteries etc.

Mercury: Fungicides, metallurgy, evaporation of Hg etc.

Chromium: Brick manufacture, stainless steel etc.

Moderately toxic

Lead: Pesticides, fertilizers, gasoline, coal, etc.

Fluorine: Pesticides, fertilizers, air pollution etc.

Nickle: Fertilizer, gasoline, batteries, coal etc.

Less toxic

Zinc: Fertilizers, pesticides, batteries, sewage waste etc.

Copper: Fungicides, fertilizers, industrial dust etc.

Boron: Fertilizers, gaseline, irrigation water etc.

C. Contamination of Soil due to acid rain

Nitrogen and sulphur dioxides liberate from motor vehicles and industries. Nitrogen and sulphur dissolve in the water vapour of atmosphere and form nitric and sulphuric acid. There help acidity rain water, which falls on the soil as acid rain. The pH of acid rain may be around 4.0.

D. Contamination of soil due to organic waste disposal

The wastes include garbage, paper products and non-biodegradable materials like polythene, glass and metals. Leaching and runoff from these sites can contaminate both surface and ground water. Contaminates include heavy metals as well as soluble and biodegradable organic materials. Many organic waste can improve nutrients to the plants. However, when excess quantities are applied, nitrate accumulated in the soil and leach into ground water which may cause nitrate toxicity.

E. Contamination of soil with salts: When the concentration of soluble salts increases to high level then it produces toxic effect directly to plants.

Detrimental effects of the toxic chemical: Toxic chemicals which are not readily biodegradable, persist in the soil or water. Non-target beneficial insects and some soil micro and macro-organisms are also affected. Soil flora and fauna may be adversely affected, as may be fish and other wild life. This problem is further compounded by entering of toxic chemicals in the food chain. For example, plant residue containing toxic materials are consumed by soil microorganism such as earthworm, the chemicals tend to concentrate in the earthworm bodies. When birds and fishes eat the earthworm, the toxic materials can build up further to lethal levels and consequently, pesticides and toxic materials can have devastating environmental pollution.

Further more, toxic elements can become part of the lifecycle of soil → plant → animal → human. Once the toxic elements become part of this cycle, they may accumulate in fish, animal and human body tissue to toxic levels.

All the pesticides are toxic and to compare the extent of their toxicity on man or animals, their median lethal doses (LD 50) is determined. It is expressed in milligrams per kilogram of body weight (mg/kg) in case of solid or liquid pesticides and for fumigants as the vapour concentration in milligrams per litre multiplied by the duration of exposure (milligram – minutes/litre). Median lethal dose (LD 50) of the pesticide is based on getting 50 of kill of the organism.

Toxicology characteristics of pesticides

Chemical	Approximate LD 50 (In mg/kg)
Insecticide	
Fluorides	50 to 100
DDT	200 to 250
BHC	600 to 1000
Endrin	10 to 12
Lindane	125 to 140
Aldrin	40 to 70
Malathion	1400 to 5800
Parathion	8 to 12
Phosdrin	7 to 10
Rogar	245 to 250
Sevin	500 to 700
Nicotine	50 to 60
Pyrethrum products	1500
Fungicides	
Organo-mercurials	50
Formalin	100
Weedicide	
Chlorates	200

Soil-pesticide interaction

Pesticide are applied to plant foliage and to soil surface or incorporated into the soil. These pesticides have six possible fates.

(i) Volatilization: Soil fumigants among insecticides are highly volatile. Few herbicides (e.g., trifluraline) and fungicides (e.g., PCNB) are highly volatile. The pesticides are lost from the soil through vaporization.

(ii) Leaching: Leaching of pesticides depends upon solubility of pesticides and their potential of adsorption with the soil colloids. In general, herbicides are more mobile than either insecticides or fungicides. Weakly asorbed herbicides are more susceptible to leaching in the soil than those that are more tightly adsorbed.

The degree of adsorption of certain herbicides

Herbicide	Adsoption to soil colloids
Dalapon	None
Bentazon	Weak
2, 4-D	Moderate
Atrazine	Strong
Paraquat	Very strong

(iii) Adsorption: Some pesticides are adsorbed by soil. The major adsorbing compounds *viz.*, the silicates and organic matter, are both involved in the adsorption reaction. In general, the order of adsorption of various pesticides is: montmorillonite > illite > kaolinite. A fine-textured soil adsorbs more than a coarse-textured soil having the same dominant clay mineral due to large surface area. Adsorbed pesticides are less liable to be washed out into the water system and pollute it.

(iv) Chemical modification: Chemical modifications help in the degradation of pesticides in the soil. Some pesticides when come in contact with soil, undergo chemical changes independent of soil microorganism. Photo decomposition of pesticides occur due to solar radiation. Other mechanism of chemical changes may be hydrolysis.

(v) Microbial metabolism: Soil microorganism is responsible for major chemical degradation of pesticides. Chlorinated hydrocarbon insecticides such as DDT, deldrin, aldrin etc. are slowly broken down in soil. Thus, these insecticides may persist in soil for 2–5 years. Organophosphate insecticides (e.g., malathion, parathion) persist in soil, only for 1–12 weeks because they are rapidly degraded by a different organisms. Risk of environmental pollution is highest with the chemicals having greatest persistence.

Persistence of some pesticides in the soil

Pesticide	Persistence in soil
Insecticides	
Arsenic	Indefinite
DDT, Aldrin, Deldrin	2–15 Year
Parathion, Malathion	1–12 Week
Sevin, carbofuran	1–8 week
Weedicides	
Atrazine, Simazine	1–2 Year
Monuron, Diuron	2–10 Months
2, 4-D, MCPA	1–5 Months
Trillate, Siramate	2–8 Weeks

Prevention and control of soil pollution

There are following methods of preventing soil. Contamination by toxic organic compounds:

1. Soil application of toxins through pesticides, fertilizers, irrigation water and solid waste should be reduced or eliminated.
2. Leaching of some toxic elements (e.g., boron) may be effective in removing excess boron from the soil.
3. Fertilizers should be applied at a time when the plants can readily absorb the nutrients.
4. Forage crops should be harvested at the stage of maturity when the concentration of the toxin in lowest.

5. Reclamation of highly salt affected soil.
6. Application of heavy phosphate in the soil reduces the availability of toxic cations but may have opposite effect on arsenic (in ion form).
7. The soil becomes reservoir for the toxins, immobilization of toxins in the soil would check their availability to the plants. Most of the toxins becomes less mobile and less available if the soil pH is neutral (7.0). Heavy metals such as copper, zinc, lead and cadmium showed highest adsorption in the soil colloids at around pH 7.0.
8. The draining of wet acid soils should be beneficial since the oxidized forms of elements such as NO_3^-, SO_4^{--}, Fe^{+++}, Mn^{+++} are readily oxidized and available to plants. This is because the reduced form of iron (Fe^{++}) and manganese (Mn^{++}) are some times present at toxic levels in wet acid soil.
9. There are differences in the ability of crops/varieties to extract toxins from soil. Sorghum absorbs more cadmium and nickle than rye and maize.

Water Pollution

Water pollution is the contamination of water bodies (e.g., lakes, rivers, oceans, aquifers and groundwater. This form of environmental degradation occurs when pollutants are directly or indirectly discharged into water bodies without adequate treatment to remove harmful compounds. Water pollution affects the entire biosphere of plants and organisms living in there water bodies, as well as organisms and plants that might be exposed to the water. In almost all cases the effect is damaging not only to individual species and populations, but also to the natural biological communities.

Water pollution is a major global problem which requires ongoing evaluation and revision of water resource policy at all levels. It has been suggested that water pollution is the leading world wise cause of deaths and diseases and that it accounts for the deaths of more than 14000 peoples daily. About 580 peoples in India die of water pollution related illness everyday. In addition to the acute problems of water pollution in developing countries, developed countries also continue to struggle with pollution problems.

Water is typically referred to as polluted when it is impaired by anthropomorphic contaminates and either does not support a human use, such as drinking water, or undergoes a marked shift in its ability to support its constituent biotic communities, such as fish. Natural phenomena such as volcanoes, algae blooms, storms and earth quakes also cause major changes in water quality and the ecological status of water.

38. QUALITY OF FERTILIZER

The quality of a fertilizer is of huge significance in India where 70% of people are employed in agriculture and on its output depends not only economic prosperity but also the livelihood of millions of Indians. The elaborate quality control mechanism that exists in India ensures identification of producer and seller of fertilizers, specifies types of fertilizer that could be manufactured and sold and arranges for periodic and surprise testing of fertilizers in markets in the special approved laboratories. Nonetheless, concerns have been raised from government and stake holders about the quality of fertilizers being sold in the market, emphasizing the need for revamping the quality mechanism from time to time.

Based on the data provided by the apex institution of Government of India for fertilizers quality- CFQCTI, its reports, data collected through a series of RTI application by FISME from Central and State laboratories and the feedback received from the members of Micronutrient Manufacturers Association, the issue of existing quality control dispensation for fertilizers has been analyzed in this report, covering three broad areas: Product and Testing Standards, Testing infrastructure and Administrative mechanism including legal provisions.

We find that there are critical deficiencies in specifications of fertilizers particularly with in micronutrient fertilizers and their mixtures. In some cases these specifications are so rigid that they in fact act against the interests of farmers as some eminently desirable elements are also grouped as impurities. According to the Government's own experts, most of these specifications are arrived at on ad-hoc basis without any detailed field study. The method of testing prescribed under standards has been questioned by experts and is not in consonance with the testing infrastructure available in testing laboratories.

The status of testing infrastructure is also a great cause of serious concern. Not only has it been found to be short of requirement, but is also being managed inefficiently. More importantly, serious concerns have been raised by experts about the level of education and training of analysts at testing laboratories. None of these laboratories is accredited by the NABL to ensure compliance of minimum standards and upkeep of testing equipment. The testing infrastructure looks to be the softest belly of the quality control mechanism and does not induce trust.

Passes in the second round. This clearly indicates that the test results are tampered with at some levels, i.e., either at the first round or the second. And there are no evidences to prove this manipulation. This needs to be checked. There should be an efficient check on the fertilizer inspectors who are involved in testing the samples.

- **Transparent system of re-testing:** Detailed information should be easily accessible on the second round testing, for instance, reasons for passing in the second round etc. Ideally, retesting should be conducted in the presence of either the parties or their authorized representatives. Competent persons may be approved to be authorized representatives of the parties. Such competent persons could be lecturers, professors of chemistry department of science or agriculture colleges and universities. In addition, for empowering the consumers, right to test or retest by the parties other than in question should be given.
- **Pragmatic tolerance limits under the FCO:** The ranges of tolerance limits specified by the FCO are not only too stringent but also impractical. The tolerance limits need to be moderate and pragmatic. For instance, the fertilizer samples are usually packed in jute bags while being transported to the laboratories for testing and there is every chance for the moisture content to rise or fall by some

points. In this case, even if the fertilizer sample meets all other prescribed limits, it would be classified as a sub standard sample.

- **Making distinction between fraud and negligence:** There is a need to realize the distinction between mistake/ negligence and fraud, i.e. between sub-standard and adulterated fertilizers. The punishment should be more intense in case of adulterated fertilizers- when there is intent to adulterate for economic benefit than the sub standard ones where minor variation occurs in chemical or physical composition due to negligence.
- **Adequate number of full time fertilizer inspectors:** There should be an adequate number of full time fertilizer inspectors who play the lead role in the entire testing process. This will ensure effectiveness of the quality control mechanism.
- **Sufficient attendance in training programmes:** The training programmes are meant for equipping the fertilizer inspectors with the ability to carry out the testing process successfully. Thus, adequate attendance of the inspectors is essential to improve the skills of the inspectors.

Unfortunately, with the disputable specifications, blemished testing infrastructure on the one hand and the administration's unbridled powers for forced closure of business and imprisonment of entrepreneurs on the other hand, conditions have been created where there is ample room for deal making and rent seeking. No wonder, a large number of samples that are failed during the first testing get passed when re-tested. Prosecution rate of offenders is very low and the process is marred with delays. The setting is particularly precarious for small enterprises- chiefly operating in the area of micro nutrients, which because of their low negotiating power acquiesce silently to the demands made by the system.

POLICY RECOMMENDATION

Given the fast deteriorating quality of soil across the country leading to stagnation in food production, there is an urgent need to ensure supply of good quality fertilizers. This indicates that there has to be a strong quality control system that effectively deters production and sale of sub-standard fertilizers. From the preceding analysis of the current practices in quality control, some clear policy recommendations follow which are presented in this section. The regulatory system responsible for carrying out the provisions of the FCO is itself flawed. This in turn leads to inefficiencies in the entire quality control mechanism. Efforts should be made in making the regulatory system more efficient, which will automatically improve the functioning of the quality control mechanism.

- **Larger role for the private sector:** The Planning Commission has acknowledged the need to take recourse to the private sector, not just in spreading knowledge and training farmers in integrated nutrient management, but also in setting up testing facilities. This would need to be covered by the FCO so that the tests done by private laboratories can be upheld. This would involve accreditation of private laboratories and benchmarks that should be put in place by regulation.

- **Judicious penalties:** The penalties prescribed by the FCO need to be judicious and moderate. The strict penalties prescribed in the form of imprisonment of up to 7 years should be replaced by moderate punishment like fines. Strict penalties are likely to discourage good quality manufacturers from entering the fertilizer business.
- **Adequate and efficient laboratory facilities:** There should be a sufficient number of laboratories with adequate capacity for testing the samples. The capacity of the laboratories should be proportionate to the number of samples to be utilized. To ensure the excellence of the fertilizer quality control laboratories, it should be made mandatory to get NABL (National Accreditation Board For Testing And Calibration Laboratories) accreditation.
- **Checking deliberate manipulation of test results:** It has been observed that around 50 percent of the samples failing in the first round eventually.

39. INDIAN FERTILIZER POLICY

Fertilizers play an important role in the global agricultural economy. It's a fact that fertilizers are an essential factor in increasing food production globally. Indian fertilizer industry has witnessed a sharp growth since the era of green revolution (1960's). It has emerged gracefully in the last 50 years and at present ranks third in the world. It has succeeded in meeting the demand of nearly all chemical fertilizers over the years and now become an important segment of Indian economy. Presently, there are 30 large size urea manufacturing plants, 21 DAP and complex fertilizers producing units, 5 units producing low analysis straight nitrogenous fertilizers and 2 units manufacturing Ammonium Sulphate as by-product. Besides, there are about 85 medium and small-scale SSP manufacturing units in the country.

Sector-Wise, Nutrient-Wise Installed Capacity of Fertilizer Manufacturing Units in India

Sr. No.	Sector	Capacity lakh (MT)		Percentage Share	
		N	P	N	P
1	Public Sector	34.98	4.02	29.04	7.15
2	Cooperative Sector	31.66	17.03	26.29	30.31
3	Private Sector	53.78	35.14	44.67	62.54
Total		**120.47**	**56.19**	**100.00**	**100.00**

Indian fertilizer policy makers have played an important role in achieving self sufficiency in fertilizer sector. Government of India is actively involved in supporting fertilizer industry. Department of Fertilizers which comes under the ambit of Ministry of Chemicals & Fertilizers looks after overall sectoral planning and development and regulation of the industry, as well as monitoring of production pricing and distribution of the output.

The main objective of Department of Fertilizers is to ensure adequate and timely availability of fertilizers at affordable prices for maximizing agricultural production in the country. An Inter-Ministerial Committee (IMC) has been constituted with Secretary

(Fertilizers) as Chairperson and Joint Secretary level representatives of Department of Agriculture & Cooperation (DAC), Department of Expenditure (DOE), Planning Commission and Department of Agricultural Research and Education (DARE).

This Committee recommends per nutrient subsidy for <N>, <P>, <K> and <S> before the start of the financial year for decision by the Government (Department of Fertilizers). The IMC also recommends a per tonne additional subsidy on fortified subsidized fertilizers carrying secondary (other than <S>) and micro-nutrients. The Committee considers and recommends inclusion of new fertilizers under the subsidy regime based on application of manufacturers/ importers and its need appraisal by the Indian Council for Agricultural Research (ICAR), for decision by the Government.

Objectives of fertilizer policy

Fertilizer is defined as any organic or inorganic substance, natural or artificial in nature supplying one or more of the chemical elements/nutrients required for plant growth. Primary nutrients viz., nitrogen (N), phosphorus (P), and potassium (K) are most frequently required in a crop fertilization programme and are needed in the larger quantity by plants as fertilizer. So, major focus of the Indian fertilizer sector policy has been on primary (macro) nutrients.

Government of India (GoI) has declared fertilizers as an essential commodity. GoI issued the Fertilizer Control Order (FCO) under the Essential Commodities Act in 1957, which was then modified in 1985. Main objectives of the fertilizer policy includes supporting domestic fertilizer production capacity so including the parameters and/ or impurities which are/or impurities which are harmful for agriculture or human consumption, not the physical impurities.

As to insulate the country from international prices, supplying quality fertilizers at inexpensive prices, ensuring sufficient availability of fertilizer in time and unbiased distribution of fertilizers to the Indian farmers. To achieve these objectives, FCO regulates fertilizers price, sale and quality. Apart from this, FCO sets specifications of all the fertilizer products for their nutrient contents and physical parameters. FCO also provides procedures for drawing and analyzing the fertilizer samples as a quality control measure.

Policy support to fertilizer industry

Government of India is dynamically involved in supporting fertilizer industry and amendments in policies have been done time to time to achieve self sufficiency in fertilizer sector. In the year 1977, Government of India introduced the retention pricing scheme (RPS) for fertilizer units. Under RPS, the difference between retention price (cost of production as assessed by the government plus 12% post tax return on networth) and the statutorily notified sale price was paid as subsidy to each unit. RPS resulted into extraordinary increase in domestic capacity/production and consumption of fertilizers. Increase in fertilizer use led to significant increase in productivity of cereals and thereby overall food grains production.

In August 1992, phosphatic and potassic fertilizers were decontrolled and the RPS covering these fertilizers was abolished. However, w.e.f. 1.10.1992, these were covered by a scheme of uniform concession. Initially, the ad-hoc Concession Scheme was introduced for subsidy on DAP, MOP, NPK Complex fertilizers. This scheme was also extended to SSP from 1993-94. Under this scheme, concession was disbursed to the manufacturers/importers by the State Governments based on the grants provided by Department of Agriculture & Cooperation (DAC). During 1997-98, DAC also started indicating an all India uniform Maximum Retail Price (MRP) for DAP/NPK/MOP.

The urea segment were continued to be under control and covered by RPS. The Government introduced a new methodology for working out subsidy to complex fertilizers w.e.f. 1.4.2002 based on the recommendations of the Tariff Commission.

In the year 2000, The Expenditure Reforms Commission (ERC) recommended inter-alia, dismantling of existing RPS for urea. Accordingly, RPS for urea units was replaced by New Pricing Scheme (NPS) in the year 2003. It aimed at inducing the urea units to achieve internationally competitive levels of efficiency, greater transparency and simplification in subsidy administration.

For ensuring Nation's food security and balanced application of fertilizers, the Government introduced Nutrient Based Subsidy (NBS) Policy for phosphatic and potassic fertilizers w.e.f. 1.4.2010. Di Ammonium Phosphate (DAP, 18-46-0), Muriate of Potash (MOP), Mono Ammonium Phosphate (MAP, 11-52-0), Triple Super Phosphate (TSP, 0-46-0), 12 grades of complex fertilizers and Ammonium Sulphate are covered under this policy.

Transportation plays vital role in fertilizer movement throughout the country. Considering its importance, NBS policy covers freight subsidy. Equated freight subsidy is paid to the manufacturers of controlled fertilizers to cover the cost of transportation from the production plants to the consumption centers. The manufacturing / importers (except ssp) have been allowed to claim freight based on the railway receipts.

Quality control policy

Quality check is an unavoidable step in selling fertilizers in India. As per the FCO norms, manufacturers/importers can sell the fertilizers to the farmers when they meet the standard of quality mentioned in the FCO order. State Governments control the quality of fertilizers supplied by the manufacturers/importers as prescribed under the FCO.

For checking the quality and issuing the certificate, there are about 74 fertilizer testing laboratories in the country. Out of which, four laboratories are working under the Government of India. These are situated at Faridabad, Kalyani, Mumbai and Chennai with an annual analyzing capacity of 134 thousand samples. The quality of the fertilizers imported in the country is invariably checked by the fertilizer quality control laboratories of the Government of India.

The State Governments are authorized to draw fertilizer samples anywhere in the country and take appropriate action against the sellers of non-standard fertilizers. Along with the cancellation of authorization certificate, strict provision includes prosecution of offenders and if convicted, sentence up to seven years imprisonment under the Essential Commodities Act.

Pricing policy

For sustained agricultural growth and to promote balanced nutrient application, it is imperative that fertilizers are made available to farmers at affordable prices. With this objective, urea being the only controlled fertilizer, is sold at statutory notified uniform sale price, and decontrolled phosphatic and potassic fertilizes are sold at indicative maximum retail prices (MRPs). Government of India can fix maximum retail prices (MRP) of all the fertilizers.

The statutorily notified sale price and indicative MRP is generally less than the cost of production of the irrespective manufacturing unit. Difference between cost of production and selling price/MRP is paid as subsidy/concession to manufacturers by the government. In the last decade, amount of fertilizer subsidy issued by the Government of India was 12695.02 crore in 2001-02, which was increased upto 64032.29 crore in 2009-10. Around Rs. 112,664 crore was budgeted for fertilizer subsidy between 2010 and 2012-13, and the revised estimates stood at Rs 128,441 crore, which is 15,777 crore higher (around 14%).

For the current fiscal year (2013-14) the government had approved a subsidy of Rs 41, 158.8585 crore for urea and Rs 29, 426.88 crore for P&K fertilizers. Recently (in 2013-14), the Government has issued Rs 70,586 crore for fertilizer subsidy against a demand of Rs 105,497 crore the Department of Fertilizers.

Import policy

Now a days, India is showing interest in importing urea. It would not only lead to minimize the demand-supply gap, but cheaper urea imports could also help to reduce subsidy burden on the government (if domestic demand of urea remains constant). India's domestic production of urea is about 220 lakh tonnes. To meet the demand, the country had imported 80.44 lakh tones urea in the entire 2012-13. Recently, India has imported about 11.09 lakh tonnes urea during April-June 2013. Average cost of imported urea was about $ 340 per tonne.

In case of P&K fertilizers like DAP and SSP, though these fertilizers are being produced in the country, the country is almost dependent on imports of the raw materials/intermediates or imports of finished phosphatic fertilizers. There are no exploitable reserves of potash in India and the country is fully dependent on its import to meet the demand of potassic fertilizers.

However, Government has taken initiatives to encourage indigenous production in P&K fertilizer sector by reducing the custom duty on phosphoric acid. P&K manufacturers in the country now can procure this important input at reasonable price. The Nutrient Based Subsidy (NBS) scheme has been announced on P & K fertilizers w.e.f. 01.04.2010 to ensure subsidy on indigenous P&K fertilizers at par with imported P&K fertilizers.

Government is also encouraging private sector and public sector companies to explore the possibilities for joint ventures abroad. It would help in ensuring uninterrupted supply of fertilizer inputs to P & K sector. During last three years, phosphatic and potash fertilizers are imported in India from various countries viz., Australia, Bahrain, Belarus, China, Canada, CIS, Estonia, Germany, Indonesia, Iran, Israel, Jordan, Korea, Kuwait, Latvia, Lithuania, Mexico, Morocco, Philippines, Russia, S. Arabia, S. Africa, Singapore, Spain, Turkey, Tunisia, USA, Ukraine and Vietnam.

Import of all fertilizers except urea is free and importers are importing these fertilizers under Open General License (OGL) as per their requirements. Import of urea in the country is restricted and permitted through three State Trading Enterprises i.e. MMTC Limited (Minerals & Metals Trading Corporation), State Trading Corporation of India Limited and Indian Potash Limited.

P&K fertilizers in India imported (LMT) during 2010–2013

Year	DAP	MAP	TSP	NPK	MOP(Agri)
2010-11	74.11	1.88	0.98	9.81	45.00
2011-12	69.05	4.94	1.60	36.73	26.94
2012-13	57.02	1.52	0.00	4.05	18.80

Indian companies who imported potassic & phosphatic- (P & K) fertilizers during last three years are Agrigold Organics Pvt. Ltd, Coromandel International Limited, Chambal Fertilizers & Chemicals Ltd., Deepak Fertilizers & Petrochemicals Ltd., Foliage, Green Star Fertilizer Ltd., Madras Fertilizers Limited, Gujarat Narmada Valley Fertilizers & Chemicals Ltd., Indian potash Limited, Gujarat State Fertilizer & Chemicals Ltd., HPM Chemicals & Fertilizes, Indian Farmers Fertilizers Cooperative Ltd., Indo Gulf Fertilizers Ltd, Mosaic India (P) Ltd., Mangalore Chemicals & Fertilizers Limited, Paradip Phosphate Ltd., Rashtriya Chemicals & Fertilizers Ltd., Sunfert International Pvt. Ltd, Tata Chemicals Ltd., KPR Fertilisers Ltd., Fertilizers & Chemicals Travancore Ltd., Shriram Fertilizers, Nagarjuna Fertilizers & Chemicals Ltd., National Fertilizers Limited, Krishak Bharati Cooperative Ltd., Zuari Holdings Limited.

Future scenario

According to the report of " The Working Group On Fertilizer Industry for the Twelfth Plan (2012-13 To 2016-17)" published by Department of Fertilizers (GoI), the demand forecasts for 'N', 'P' and 'K' are estimated at 200.35 LMT, 96.00 LMT and 41.74 LMT respectively in the terminal year (2016-17) of the Twelfth Five Year Plan. This corresponds to a demand of 336.77 LMT for urea, 124.13 LMT for DAP, 47.93 LMT for MOP, 114.20 LMT for complex fertilizers and 59.48 LMT for SSP.

World fertilizer use in 2025 is projected to be 185 mmt, composed of 107 mmt of nitrogen (N) fertilizers, 43 mmt of phosphorous (P), and 35 mmt of potassium (K).

China, India, the U.S., and the EU-27 countries account for more than two-thirds (65%) of the world's fertilizer consumption in agriculture. India is the third-largest fertilizer consumer. India's projected fertilizer use in 2025 is 15.37 mmt of N, 5.78 mmt of P, and 2.75 mmt of K.

As of now, the country has achieved 75% self-sufficiency in production capacity of urea with the result that India could substantially manage its requirement of nitrogenous fertilizers through the indigenous industry and imports. Similarly, 50% indigenous capacity has developed in respect of phosphatic fertilizers to meet domestic requirements. However, the raw materials and intermediates for the same are largely important.

To meet the projected fertilizer demand and for ensuring a reasonably high level of self-sufficiency, a favorable and stable policy environment is an essential criterion. Besides this, factors like additional capacity, technology support, raw materials availability, transportation, price incentives, credit availability, research and extension support and regulatory frameworks would play important role in future growth and achieving stability of the fertilizer industry in India.

40. THE FERTILIZER (CONTROL), ORDER 1985

I. Introduction

G.S.R. 758 (E) in exercise of the powers conferred by section 3 of the Essential Commodities Act. 1955 (10 of 1955), the Central government Hereby makes the following Order, Namely:

1. Short title and commencement

(1) This order may be called the Fertilizer (Control) Order, 1985.

(2) It shall come into force on the date of its publication in the official Gazette.

2. Definitions

In this order, unless the context otherwise requires:

(a) "Act" means the Essential Commodities Act, 1955 (10 of 1955)

(b) "Certificate of source" means a certificate given by a State Government, Commodity Board, manufacturer, importer, pool handling agency or — as the case may be, wholesale dealer indicating therein the source from which Fertilizer for purpose of sale is obtained.

(c) "Commodity Board" means the Coffee Board constituted under section 4 of the Coffee Act, 1942 (7 of 1942) or the Rubber Board constituted under section 4 of the Rubber Act 1947 (24 of 1947) or the Tea Board constituted under section 4 of the Tea Act, 1953 (29 of 1953) or as the case may be, the Cardamom Board constituted under section 4 of the cardamom Act, 1965 (42 of 1965)

(d) Compound or complex Fertilizer" means a Fertilizer containing two or more nutrients during the production of which chemical reaction takes place.

(e) "Controller" Means the person appointed as controller of Fertilizers by the Central Government and includes any other person empowered by the Central

Government to exercise or perform all or any of the powers, or as the case may be, functions of the Controller under this order.

(f) "Dealer" means a person carrying on the business of selling Fertilizers whether wholesale or retail or industrial use and includes a manufacturer, importer, and a pool handling agency carrying on such business and the agents of such person, manufacture, importer or pool handling agency.

(g) Clause 'g' deleted vide S.O. 725 (E) dated (E) dated 28.7.88

(h) "Fertilizer" means any substance used or intended to be used as a Fertilizer of the soil and/or crop and specified in Part A of Schedule-1 and includes a mixtures of Fertilizers and special mixture of Fertilizers;

(i) "Form" means a form appended to this Order;

(j) "Grade" means the nutrient element contents in the Fertilizer expressed in percentage;

(k) "Granulated mixture" means a mixture of Fertilizers made by intimately mixing two or more Fertilizers with or without inert material and granulating the together, without involving any chemical reaction;

(l) "Importer" means a person who imports Fertilizer in accordance with the Export and import policy of the Central Government, as amended from time to time.

  (i) "Inspector" means as inspector of Fertilizers appointed under clause 27;

  (ii) "Industrial dealer" means a dealer who sells Fertilizers for industrial purposes;

  (iii) "Industrial purposes" means the use of Fertilizer for purposes other than fertilization of soil and increasing productivity of crops;

(m) "Manufacturer" means a person who produces Fertilizers or mixtures of Fertilizers and the expression manufacture" with its grammatical variations shall be construed accordingly,

(n) "Mixture of Fertilizer" means a mixture of fertilizers made by physical mixing two or more Fertilizers with or without inert material in physical or granular form and includes a mixture of I\IPK Fertilizers, a mixture of micro nutrient Fertilizers and a mixture of NPK with micro nutrient Fertilizers.

(o) "Offer for sale" includes a reference to an intimation by a person of a proposal by him for the sale of any Fertilizer, made by publication of a price list, by exposing the Fertilizer for sale indicating the price, by furnishing of a quotation of otherwise howsoever;

(p) "Physical mixture" means a mixture of Fertilizer made by physical mixing of two or more Fertilizers with or without inert material necessary to make a required grade, without involving any chemical reaction;

(q) "Prescribed standard" means:

    (i) In relation to a Fertilizer included in column 1 of part A of Schedule -I the standard set out in the corresponding entry in column, 2, subject to the limits of permissible variation as specified in part B of that Schedule and

(ii) In relation to mixture of Fertilizer the standard set out in respect of that mixture under sub-clause(1) of clause 13 by the Central Government, subject to the limits of permissible variation as specified in part B of Schedule-I.

(iii) In relation to mixture of Fertilizers, standard set out in respect of that mixture under sub-clause (2) of clause 13 by the State Government, subject to limits of permissible variation as specified in Part B of Schedule-I

(r) "Pool handling agency" means an agency entrusted by the Central Government with functions relating to handling and distribution of imported Fertilizer;

(s) "Registering authority" means a registering authority appointed under clause 26;

(t) "Retail dealer" means a dealer who sells Fertilizer to farmers or plantations for agricultural use such as for fertilization of soil and increasing productivity of crops;

(u) "Schedule " means a Schedule appended to this Order;

(v) "Special mixture of Fertilizers" means any mixture of Fertilizers prepared for experimental purposes in pursuance of a requisition made by any person (including a person engaged in the cultivation of tea, coffee or rubber) for sale to that person in such quantity and within such period as may be specified in such requisition and

(w) "Wholesale dealer" means a dealer who sell Fertilizer otherwise than in retail" for agricultural use such as for fertilization of soil and increasing productivity of crops.

II. PRICE CONTROL

3. Fixation of prices of Fertilizers.

(1) The Central Government may, with a view to regulating equitable distribution of Fertilizers and making Fertilizers available at fair prices, by notification in the official Gazette, fix the maximum prices or rates at which any Fertilizer may be sold by a dealer, manufacturer, importer or a pool handling agency.

(2) The Central Government may having regard to the local conditions of any area, the period of storage of Fertilizer and other relevant circumstances, fix different price or rates for Fertilizer having different periods of storage or for different areas or for different classes of consumers.

(3) No dealer, manufacturer importer or pool handling agency shall sell or offer for sale any Fertilizer at a price exceeding the maximum price or rate fixed under this clause.

4. Display of stock position and price list of Fertilizer.

Every dealer, who makes or offers to make a retail sale of any Fertilizers, shall prominently display in his place of business:-

(a) The quantities of opening stock of different Fertilizers held by him on each day.

Explanation: The actual stocks at any point of time during the day may be different from that of the displayed opening stocks to the extent of sale and receipt of such Fertilizers up to the time of inspection during that day.

(b) A list of prices or rates of such Fertilizers fixed under clause 3 and for the time being in force.

5. Issue of cash/credit memorandum

Every dealer shall issue a cash or credit memorandum to a purchaser of a Fertilizer in Form M:

III. CONTROL ON DISTRIBUTION OF FERTILIZERS BY MANUFACTURER/IMPORTER

6. **Allocation of Fertilizers to various States.**

The Central Government may, with a view to securing equitable distribution and availability of Fertilizers to the farmers in time, by notification in the official Gazette, direct any manufacture/importer to sell the Fertilizers produced by him in such quantities and in such State or States and within such period as may be specified in the said notification.

IV. REGISTRATION OF DEALER

7. **Dealers to be registered:** No person, including a manufacturer, an importer, a pool handling agency, a wholesale dealer, a retail dealer and an industrial dealer shall offer for sale or carry on the business of selling Fertilizers at any place except under and in accordance with the terms and conditions of a certificate of registration granted to him under clause 9.

   Provided that a State Government may, by notification in the official Gazette, exempt from the provisions of this clause any person selling Fertilizers to farmers in such areas and subject to such conditions as may be specified in that notification.

8. **Application for registration:** Every person desiring to obtain a certificate of registration under this Order for selling Fertilizers, whether in wholesale or retail or both shall make an application for registration to the registering authority or, as the case may be, for industrial use thorough registered post to the controller in Form A, in duplicate, together with the fee prescribed under clause 36 and a certificate of source, "in Form O".

   Provided that where the applicant is a State Government, a manufacturer or an importer or a pool handling agency, it is not necessary to enclose a certificate of source along with the application.

   Provided further that separate certificate of registration shall be obtained for selling Fertilizer in wholesale and retail and in case of a State Government, a manufacturer, importer and a pool handling agency, also for selling Fertilizer for industrial use;

Provided also that where Fertilizers are obtained for sale from different sources, a certificate or source from each such source shall be furnished.

Provided also that for obtaining a certificate of registration for industrial dealer, a certificate of source obtained only from the State Government or a manufacturer or an importer or pool handling agency shall be furnished. (Reference 459(E) dated 22 may 1995)

9. **Grant or refusal of certificate or registration:** The registration authority or, as the case may be, the Controller, shall grant a certificate of registration in form 'B' within thirty days of the receipt of application to any person who applies for it under clause 8;

Provided that no certificate of registration shall be granted to, A person,

(a) If his previous certificate of registration is under suspension ; or

(b) If his previous certificate of registration has been cancelled within a period of one year immediately preceding the date of application; or

(c) If he has been convicted of an offence under the Act, or any Order made there under within three years immediately preceding the date of making the application; or

(d) If he fails to enclose with the application a certificate of source; or

(e) If the application is incomplete in any respect; or

(f) If he makes an application for obtaining the certificate of registration for If dustrial dealer and excepting if he is a manufacturer, importer or pool handling agency, holds a valid certificate of registration for wholesale dealer or retail dealer or both, and as the case may be, the vice-versa.

10. **Period of validity of certificate of registration:** Every certificate of registration granted under clause 9 or as the case may be renewed under clause 11 shall, unless suspended or cancelled, be valid for a maximum period of three years, from the date of issue.

11. **Renewal of certificate of registration-**

(1) Every holder of a certificate of registration desiring to renew the certificate of registration granted under clause 9 shall, before the date of expiry of such certificate of registration make an application for renewal to the registering authority or as the case may be, the Controller, in Form C, in duplicate, together with the fee prescribed under clause 36 for such renewal and a certificate of source as required under clause 8.

(2) On receipt of such application, together with such fee and certificate of source, the registering authority or as the case may be, the Controller, may renew the certificate of registration;

Provided that a certificate of registration shall not be renewed if the holder of the certificate of registration did not sell any Fertilizer during the period of one year immediately preceding the date of expiry of the period of validity of the certificate of registration sought to be renewed.

(3) If any application for renewal is not made before the expiry of the period of validity of the certificate of registration but is made within one month from the date of such expiry, the certificate of registration may be renewed on payment of such additional fee as may be prescribed by the State Government or as the case may be, the Controller, in addition to the fee for renewal of the certificate of registration.

(4) Where the application for renewal is made within the time specified in sub clause (1) or sub-clause (3) the application shall be deemed to have held a valid certificate of registration, until such date as the registering authority or, as the case may be, the controller, passes orders on the application for renewal.

(5) If an application for renewal of a Certificate of registration is not made within one month of the date of expiry of the period of validity of the certificate of registration, the certificate of registration shall be deemed to have lapsed on the date on which its validity expired and any business carried on after that date shall be deemed to have been carried on in contravention of clause 7.

V. MANUFACTURE OF MIXTURE OF FERTILIZERS

12. **Registration on preparation of mixture of Fertilizers:** No person shall carry on the business of preparing any mixture of Fertilizers, or special mixture of Fertilizer except under and in accordance with the terms and conditions of a certificate of manufacture granted to him under clauses 15 or 16.

13. **Standards of mixtures of Fertilizers-**

(1) Subject to the other provisions of the order, no person shall manufacture any mixture of Fertilizer whether of solid or liquid Fertilizer unless such mixture conforms to the standards set out in the notification to be issued by the central Government in the official Gazette.

(2) Subject to the other provisions of this order, no person shall manufacture any mixture of Fertilizer except of liquid Fertilizer unless such mixture conforms to the standards set out in the notification to be issued by the State Government in the Official Gazette;

(3) Notwithstanding anything contained in this order, those units which are already manufacturing mixtures of Fertilizers, as on the date of publication of this Order, for which standards have not yet been specified under clause 13(2) may continue to manufacture for sale, sell, offer for sale, stock or exhibit for sale or distribute such mixture of micro nutrient Fertilizer for a period up to 27th July, 1992 from such date or till the standards are prescribed by the State Government whichever is earlier subject to the condition that before the expiry of said period, they will follow the standards prescribed by State Government and obtain Certificate of Manufacture under clause 15.

(4) No Certificate of manufacture shall be granted in respect of any Fertilizer which does not confirm to the standards set out in the notification refereed in sub clause (1) or (2)

14. Application for certificate of manufacture of mixtures of Fertilizers

(1) Every person desiring to obtain a certificate of manufacture for preparation of any mixture of Fertilizers of special mixture of Fertilizers shall possess such mixture, and possess the minimum laboratory facility as specified in clause 21A of this Order.

(2) An applicant for a certificate of manufacture for preparation of mixture of Fertilizers or special mixture of Fertilizers shall make an application to the registering authority.

(a) If he is an applicant for a certificate of manufacture for any mixture of Fertilizers in Form D, in duplicate, together with the fee prescribed therefore under clause 36;or

(b) If he is an applicant for a certificate of manufacture for any special mixture, in Form E, in duplicate, together with the fee prescribed therefore under the said clause 36 and an attested copy of the requisition of the purchaser.

15. Grant or refusal of certificate of manufacture for preparation of mixture of Fertilizers

(1) The registering authority may, for reasons to be recorded in writing, refuse to grant a certificate of manufacture in respect of any mixture of Fertilizers or special mixture of Fertilizers and shall furnish to the applicant a copy of the order so passed.

(2) Where an application for a certificate of manufacture for mixture of Fertilizers is not refused under sub-clause (1) the registering authority shall grant a certificate of manufacture in Form F and where an application for a certificate of manufacture for a special mixture is not refused under that sub-clause, such authority shall grant a certificate of manufacture to the applicant in Form G,

16. Conditions for grant of certificate of manufacture in respect of special mixture of Fertilizers and period of validity of such certificate

(1) No certificate of manufacture in respect of any special mixture of Fertilizers shall be granted to an applicant unless he holds a valid certificate of manufacture under this order for any mixture of Fertilizers.

(2) Every certificate of manufacture granted in respect of any special mixture of Fertilizers shall be valid for a period of three months from the date of its issue.

Provided that the registering authority may, if it is satisfied that it is necessary so to do, extend the said period to such further period or periods as it may deem fit, so however, that the total period or period so extended shall not exceed six months.

17. Period of validity of a certificate of manufacture for preparation of mixtures of Fertilizers.

Every certificate of manufacture granted under clause 15 for preparation of a mixture of Fertilizers shall, unless suspended or cancelled, be valid for a period of three years from the date of issue.

18. Renewal of certificate of manufacture for preparation of mixtures of Fertilizers.

(1) Every holder of a certificate of manufacture for preparation of mixture of Fertilizers desiring to renew the certificate, shall, before the date of expiry of the said certificate of manufacture make an application to the registering authority in Form D in duplicate, together with the fee prescribed for this purpose under clause 36.

(2) On receipt of an application for renewal as provided in sub-clause (1) and keeping in view the performance of the applicant and other relevant circumstances, the registering authority may, if he so decides, renew the certificate of registration by endorsement on Form F and in case the certificate of registration is not renewed, the registering authority shall record in writing his reasons for not renewing the certificate of registration.

(3) If an application for renewal is not made before the expiry of the certificate of registration but is made within one month from the date of expiry of the certificate of registration but is made within one month from the date of expiry of the certificate of registration, the certificate of registration may be renewed on payment of such additional fee as may be prescribed by the State Government for this purpose.

(4) Where the applicant for renewal is made within the time specified in sub clause (1) or Sub-Clause (3) the applicant shall be deemed to have held a valid certificate of registration until such date as the registering authority passes order on the application for renewal.

(5) If an application for renewal of a certificate of manufacture is not made within the period stipulated under sub-clause (1) or, as the case may be, under Sub-clause (3) the certificate of manufacture shall be deemed to have expired immediately on the expiry of its validity period, and any business carried on after that date shall be deemed to have been carried on in contravention of clause 12.

VI. RESTRICTIONS ON MANUFACTURE/IMPORT, SALE, ETC. OF FERTILIZER

19. Restriction on manufacture/import, sale and distribution of Fertilizers.

No person shall himself or by any other person on his behalf :-

(a) Manufacture/import for sale, sell, offer for sale, stock or exhibit for sale or distribute any Fertilizer which is not of prescribed standard.

(b) Manufacture/import for sale, sell, offer for sale, stock or exhibit for sale or distribute any mixture of Fertilizers, which is not of prescribed standard (Subject to such limits of permissible variation as may be specified from time to time by the Central Government) or special mixture of Fertilizers which does not conform to the particulars specified in the certificate of manufacture granted to him under this Order in respect of such special mixture.

(c) Sell, offer for sale, stock or exhibit for sale or distribute:-

(i) Any Fertilizer the container where of is not packed and marked in the manner laid down in this Order,

(ii) Any Fertilizer which is an imitation or a substitute for another Fertilizer under the name of which it is sold,

(iii) Any Fertilizer which is adulterated.

Explanation: A Fertilizer shall be deemed to be adulterated, it it contains any substance the addition of which is likely to eliminate or decrease its nutrient contents or make the Fertilizer not conforming to the prescribed standard.

(iv) Any Fertilizer the label or container where of bears the name of any individual firm or company purporting to be manufacturer/importer of the Fertilizer, which individual, firm or company is fictitious or does not exist;

(v) Any Fertilizer, the label or container where of or anything accompanying therewith bears any statement which makes a false claim for the Fertilizer or which is false or misleading in any material particular.

(vi) Any substance as a Fertilizer which substance is not, in fact a Fertilizer or

(vii) Any Fertilizer without exhibiting the minimum guaranteed percentage by weight of plant nutrient.

20. Specifications in respect of imported Fertilizer.

Notwithstanding anything contained in this order, the Central Government may, by an order, published in the Official Gazette, fix separate specifications in respect of imported Fertilizers.

21. Manufacturers/importers pool handling agencies to comply with certain requirement in regard to packing and marking, etc.

Every manufacturer/importer and pool handling agency shall in regard to packing and marking of containers of Fertilizers, comply with the following requirements, namely :-

(a) Every container in which any Fertilizer is packed shall bear** only such particulars and nothing else as may from time to time, be specified by the Controller in this behalf; and

(b) Every container shall be so packed and sealed that the contents thereof cannot be tampered with without breaking the seal;

Provided that where Fertilizer manufactured in India are packed in bags stitched in hand such bags shall bear lead seals, so that the contents thereof cannot be tampered with without breaking the seals; Provided further that lead sealing shall not be necessary:

(i) If such bags are machine stitched in such a manner that contents thereof cannot be tampered with without a visible break in the stitching; and

(ii) In the case of Fertilizers imported from abroad and packed in bags stitched in hand, in such a manner that the contents thereof cannot be tampered with without visible break in the stitching.

21. A Manufacturers to comply with certain requirements for laboratory facilities: Every manufacturer shall, in order to ensure quality of their product, possess minimum laboratory facility, as may be specified from time to time by the Controller.

22. Bulk sale of Fertilizers.

Notwithstanding anything contained in this Order:-

(a) A retail dealer may retain at any time one bag or container of each variety of Fertilizer in an open and unsealed condition for the purpose of sale;

(b) A Manufacturer/Importer may sell the Fertilizer manufactured/imported by him in bulk to a manufacturer of mixture of Fertilizers, compound/complex Fertilizers or special mixture of Fertilizers; and

(c) The Central Government may by notification published in the official Gazette in this behalf authorize a manufacturer/importer to sell any Fertilizer manufactured/imported by him in bulk also direct to farmers for such period as may be specified in that notification:

Provided that a certificate indicating the minimum guaranteed percentage of plant nutrients is issued by the manufacturer/importer to each -farmer at the time of such sale.

23. Disposal of non-standard Fertilizers.

(1) Notwithstanding anything contained in this order, a person may sell, offer for sale, stock or exhibit for sale or distribute, any Fertilizer which, not being ar. adulterated Fertilizer, does not conform to the prescribed standard (hereinafter in this Order referred to as non-standard Fertilizer) subject to the conditions that:-

(a) The container of such non-standard Fertilizer is conspicuously super scribed in red colour with the words "non-standard" and also with the sigh "X" and

(b) An application for the disposal of non-standard Fertilizers in Form H is suited to the registering authority to grant a certificate of authorization for sale of such Fertilizers and a certificate of authorization with regard to their disposal and price is obtained in Form I

(c) Such non-standard Fertilizer shall be sold only to the manufacturers of mixtures of Fertilizers of special mixtures of Fertilizers of research farms of Government or Universities or such bodies.

(2) The price per unit of the non-standard Fertilizer shall be fixed by the registering authority after satisfying itself that the sample taken is a representative one and after considering the nutrient contents in the sample determined on the basis of a chemical analysis of the non-standard Fertilizer.

(3) The Central Government may by notification in the official Gazette and subject to the conditions, if any, laid down in that notification and subject to guidelines issued in this regard by the Central Government exempt such pool handling agencies as it deems fit, from complying with conditions laid down in paragraphs (a) and (b) of the sub-clause(l)4

24. Manufacturers/pool handling agencies to appoint officers responsible with **compliance of the Order.**

Every manufacturing organization importer and pool handling agency shall appoint in that organization and in consultation with the Central Government, an officer, who shall be responsible for compliance with the provisions of this Order.

25. Restriction on sale / use of Fertilizers:

(1) No person shall, except with the prior permission of the Central Government and subject to such terms and conditions as may be imposed by such Government, sell or use Fertilizer, for purposes other than fertilization of soils and increasing productivity of crops.

Provided that the price of Fertilizers permitted for sale for industrial use shall be no profit no loss price, excluding all subsidies at the production, import, handling or on sale for agricultural consumers;

Provided further that wherever customs or excise duties are chargeable these may be added to the price so fixed.

Provided also that in the case of non-standard Fertilizers, reductions shall be made from the no profit no loss price, indicated above, proportionate to the loss of nutrient contents.

(2) Not withstanding anything contained in sub-clause (1) no prior permission for use of Fertilizer for industrial purposes shall be necessary when the Fertilizer certificate of registration granted under clause 9.

(3) Any person possessing a valid certificate of registration for industrial dealer, unless such person is a State Government, a manufacturer/importer or a pool handling agency, shall not carry on the business of selling Fertilizers for agricultural purposes, including a wholesale dealer or a retail dealer. However, in case of a State Government, a manufacturer or a importer or a pool handling agency possessing a valid certificate of registration for sale of Fertilizer for industrial use, and also for sale of Fertilizer for industrial use, and also for sale of Fertilizer for agricultural use, whether in wholesale or retail or both, shall not carry on the business of selling Fertilizers both for industrial use and agricultural use in the same premises.

VII. ENFORCEMENT AUTHORITIES.

26. Appointment of registering authority.

The State Government may, by notification in the Official Gazette, appoint such number of persons, as it thinks necessary, to be registering authorities for the purpose of this order excepting for grant or renewal of certificate of registration for industrial dealers, and may, in any such notification define the limits of local area within which each such registering authority shall exercise his jurisdiction.

27. Appointment of inspectors.

The State Government, or the Central Government may, by notification in the Official Gazette appoint such number of persons, as it thinks necessary, to be inspectors of Fertilizers for the purpose of this Order, and may, in any such notification, define the limits of local area within which each such inspector shall exercise his jurisdiction.

27A. Qualifications for appointment of Fertilizer inspectors.

No person shall be eligible for appointment as Fertilizer inspector under this order unless he possesses the following qualifications, namely:

(1) Graduate in agriculture or science with chemistry as one of the subjects, from a recognized university; and

(2) Training or experience in the quality control of Fertilizers and working in the State or Central Government Department of Agriculture.

28. Powers of inspectors.

(1) An inspector may, with a view to securing compliance with this order:-

(a) Require any manufacturer, importer, pool handling agency, wholesale dealer or retail dealer to give any information in his possession with respect to the manufacture, storage and disposal of any Fertilizer manufactured or in any manner handled by him.

(b) Draw samples of any Fertilizer in accordance with the procedure of drawl of samples laid down in Schedule II.

Provided that the inspector shall prepare the sampling details in duplicate in Form 3 and hand over one copy of the same to the dealer or his representative from whom the sample has been drawn;

(c) Enter upon and search any premises where any Fertilizer is manufactured/ imported or stored or exhibited for sale if he has reason to believe that any Fertilizer has been or is being manufactured/imported, sold, offered for sale, stored, exhibited for sale or distributed contrary to the provisions of this Order.

(d) Seize or detain any Fertilizer in respect of which he has reason to believe that a contravention of this Order has been or is being or is about to be committed.

(e) Seize any books of accounts or documents relating to manufacture, storage or sale of Fertilizers, etc. in respect of which he has reason to believe that any contravention of this Order has been or is being or is about to be committed;

Provided that the inspector shall give a receipt for such Fertilizers or books of accounts or documents so seized to the person from whom the same have been seized;

Provided further that the books of accounts or documents or documents so seized shall be returned to the person from whom they were seized after copies thereof or extracts therefore as certified by such person, have been taken.

(2) Subject to the proviso to paragraphs (d) and (e) of sub-clause (1) the provisions of the Code of Criminal procedure, 1973 (2 of 1974) relating to search and seizure shall, so far as may be, apply to searches and seizures under this clause.

Provided also that the inspector shall give the stop sale notice in writing to the person whose stocks have been detained and initiate appropriate action as per the provisions of this order within a period of twenty-one days. If no action has been initiated by the inspector within the said period of twenty-one days from the date of issue of the said notice, the notice of stop sale shall be deemed to have been revoked.

(3) Where any Fertilizer is seized by an inspector under this clause he shall forth with report the fact of such seizure to the collector whereupon the provisions of sections 6A 6B 6C 6D and 6E of the Act, shall apply to the authority disposal and confiscation of such Fertilizers.

(4) Every person, if so required by an inspector, shall be bound to afford all necessary facilities to him for the purpose of enabling him to exercise his powers under sub-clause (1)

VIII. ANALYSIS OF SAMPLES

29. Laboratory for analysis

A Fertilizer sample, drawn by an inspector, shall be analyzed in accordance with the instruction contained in Schedule II (Part B) in the "Central Quality Control and Training institute, Faradized or Regional Fertilizer Control Laboratories at Bombay, Madras or Kalyani (Calcutta) or in any other laboratory notified for this purpose by the State Government. [Q|

29A. Qualifications for appointment of Fertilizer analyst in the Fertilizer control laboratories.

No person shall be eligible for appointment as Fertilizer analyst for analysis of Fertilizers samples in the laboratories notified under clause 29 of the Order, unless he possesses the following qualifications, namely:

(1) Graduate in Agriculture or Science with chemistry as one of the subjects from a recognized university; and

(2) Training in Fertilizer quality control and analysis at Central Fertilizer Quality Control and Training Institute, Faridabad.

Provided that the Fertilizer analysts appointed before the commencement of this order, who do not possess the requisite training, shall under go prescribed training, within a period of three years, in the Central Fertilizer Quality Control and Training Institute, Faridabad from the date of commencement of this Order.

30. Time limit for analysis, and communication of result.

(1) Where sample of a Fertilizer has been drawn, the same shall be dispatched along with a memorandum in Form K to the laboratory for analysis within a period of seven days from the date of its drawl.

(2) The laboratory shall analyze the sample and forward the analysis report in Form L within 50 days from the date of receipt of the sample in the laboratory to the authority specified in the said memorandum.

(3) The authority to whom the analysis report is sent under sub-clause (2) shall communicate the result of the analysis to the dealer/manufacture/importer/pool handling agency from whom the sample was drawn within 30 days from the date of receipt of the analysis report of the laboratory.

IX. MISCELLANEOUS

31. Suspension/ Cancellation of Registration Certificate:-

(1) A Registering authority or, as the case may be, the Controller may, after giving the holder of a certificate of registration, of a certificate of manufacture or any other certificate granted under this Order, an opportunity of being heard, suspend or cancel such certificate on any of the following grounds, namely:-

(a) That such certificate has been obtained by misrepresentation as to material particulars;

(b) That any of the provisions of this order or any of the terms and conditions of such certificate has been contravened or not fulfilled;

Provided that while canceling the certificate the holder thereof may be allowed a period of 30 days to dispose of the balance stock of Fertilizers if any, held by him;

Provided further that the stock of Fertilizer lying with the holder after the expiry of the said 30 days period shall be confiscated;

(2) Where the contravention alleged to have been committed by a person is such as would, on being proved, justify cancellation of the certificate of registration or, as the case may be, certificate of manufacture or any other certificate granted under this Order to such person the registering authority or, as the case may be, the Controller any, without any notice, suspend such certificate as an interim measure;

Provided that the registering authority or, as the case may be, the controller shall immediately furnish to the person details and the nature of contravention alleged to have been committed by such person and, after giving the person an opportunity of being heard, pass final orders either revoking the order of suspension or canceling the certificate within fifteen days from the date of issue of the order of suspension;

Provided further that where no final rode is passed within the period as specified above, the order of interim suspension shall be deemed to have been revoked without prejudice, however, to further action which the registering authority or as the case may be, the controller may take against the holder of the certificate under sub-clause (1)

(3) Wherever a certificate is suspended or cancelled under this clause, the registering authority or, as the case may be, the Controller shall record a brief statement or

the reasons for such suspension or, as the case may be cancellation and furnish a copy thereof to the person whose certificate has been suspended or cancelled.

(4) Wherever the person alleged to have committed the contravention is an industrial dealer, the registering authority may also take action against the holder of such certificate of registration under sub-clause (1) and (2)

Provided that there such certificate is suspended or cancelled the registering authority shall, within a period of fifteen days from the date of issue of such Order of suspension or cancellation, furnish to the controller also, besides sending the same to the person whose certificate has been suspended or cancelled, a detailed report about the nature of contravention committed and a brief statement of the reasons for such suspension or, as the case may be cancellation;

Provided further that the controller, shall, in case of the order for suspension passed by the registering authority, on receipt of the detailed report and after giving the person an opportunity of being heard, pass final order either revoking the order of suspension or canceling the certificate of registration, within fifteen days from the date of receipt of the detailed report from the registering authority, failing which the order of interim suspension passed by the registering authority shall be deemed to have been revoked, without prejudice however, to further action which the controller may take against the holder of certificate under sub-clause (1)

Provided also that the order of cancellation passed by the registering authority shall remain effective as if it has been passed by the Controller till such time the Controller, on receipt of the detailed report from the registering authority, and if deemed necessary, after giving the person a fresh opportunity of being heard pass the final order either revoking or confirming the order of cancellation.

32. Appeal

Any person, "excepting an industrial dealer or, as the case may be, the person desiring to obtain a certificate of registration for industrial dealer, aggrieved by:-

(1) An order

(a) Refusing to grant, amend or renew a certificate of registration for sale of Fertilizer; or

(b) Refusing to grant a certificate of manufacture for preparation of mixture of Fertilizers or special mixture of Fertilizers, or

(c) Suspending or canceling a certificate of registration or manufacture or;

(2) Non-issuance of certificate or registration to him within the stipulated period may within sixty days from the date of receipt of such order or, as the case may be, from the date of expiry of such stipulated period, appeal to such authority as the State Government may specify in this behalf, and the decision of such authority shall be final.

33. Grant of duplicate copies of certificate of registrations, etc.

Where a certificate of registration or a certificate of manufacture or any other certificate granted or, as the case may be, renewed under this order is lost the registering authority, or, as the case may be, the controller may, on an application made in this behalf, together with the fee prescribed for this purpose under clause 36 grant a duplicate copy of such certificate.

34. Amendment of certificate of registration

The registering authority or, as the case may be, the controller may, on an application made in this behalf, together with the fee prescribed for the purpose under clause 36, amend a certificate of registration.

35. Maintenance of records and suission of returns, etc.

(1) The controller may by an order made in writing direct the dealers, manufacturers/ importers, and pool handling agencies[1].-

(a) To maintain such books of accounts, records, etc. relating to their business in Form "N" and

(b) To suit to such authority, returns and statements in such form and containing such information relating to their business and within such time as may be specified in that order.

(2) Where a person holds certificates of registration for retail sale and wholesale of Fertilizers, he shall maintain separate books of accounts for these two types of sales made by him.

(3) Where a State Government, a manufacturer, + an importer and a pool handling agency holds valid certificates of registration for sale or Fertilizers in wholesale or retail or both and also for sale for industrial use, he shall maintain separate books of accounts for these two or three types of sales made by him.

(4) Every importer shall inform the Director of Agriculture of the State in which he intends to discharge the imported Fertilizer, under intimation to the Central Government , before the import is made or within a period of fifteen days after an indent for import is placed, the following details, namely:-

(i) Name of Fertilizer,

(ii) Name of country of import,

(iii) Name of manufacturer

(iv) Quantity to be imported,

(v) Date of arrival of the consignment,

(vi) Name of the discharge port,

(viii) Other information."

36. Fees

(1) The fees payable for grant, amendment or renewal of a certificate of registration or certificate of manufacture a duplicate of such certificates or renewal there of under this Order shall be such as the State Government may, with prior approval of the controller, from time to time fix, subject to the maximum fees fixed for different purposes by the Central Government and different fees may be fixed for different purposes or for different classes of dealers or for different types of mixtures of Fertilizer or special mixture.

Provided that no fee shall be payable by a person holding a valid certificate of manufacture for preparation of mixture of Fertilizers or special mixture of Fertilizer issued under the Fertilizer (Control) Order 1957 and making an application for grant or a fresh certificate of manufacture for preparation of such mixture after the commencement of this order.

(2) The authority to whom and the manner in which the fee fixed under sub-clause (1) Shall be paid, shall be such as may be specified by the State Government by notification in the Official Gazette.

(3) Any fee paid under sub-clause (1) shall not be refundable unless the grant or renewal of any certificate of registration or certificate of manufacture or duplicate copy of such certificate or renewal under this order has been refused.

(4) The fees payable for grant, amendment, renewal or duplicate copy of certificate registration for industrial dealer and the authority to whom and the manner in which such fee shall be paid, shall be such as may be specified by the Controller from time to time by notification in the official Gazette.

37. Service of orders and directions.

Any order or direction made for issued by the controller or by any other authority under this order shall be served in the same manner as provided in sub-section (5) of section 3 of the Act.

38. Advisory Committee.

(1) The Central Government may by notification in the Official gazette and on such terms and conditions as may be specified in such notification, constitute a committee called the Central Fertilizer Committee consisting of a Chairman and not more than ten other persons having experience or knowledge in the field, who shall be members of the committee, to advise the Central Government regarding:

(i) Inclusion of a new Fertilizer, under this order;

(ii) Specifications of various Fertilizers;

(iii) Grades/formulations of physical/granulated mixtures of Fertilizers that can be allowed to be prepared in a State.

(iv) Requirements of laboratory facilities in manufacturing unit, including a unit manufacturing physical/granulated mixtures of Fertilizers.

(v) Methods of drawl and analysis of samples.

(vi) Any other matter referred by the Central Government to the Committee.

(2) The committee may, subject to the previous approval of the Central Government, made bye-laws fixing the quorum and regulating its own procedure and the conduct of all business to be transacted by it.

(3) The committee may co-opt such number of experts and for such purposes or periods as it may deem fit, but any expert so co-opted shall not have the right to vote.

(4) The committee may appoint one or more sub-committees, consisting wholly of members of the committee or wholly of co-opted members or partly of the

members of the committee and partly of co-opted members as it thinks fit, for the purpose of discharging such of its functions as may be delegated to such sub-committee or sub-committees by the Central Fertilizer Committee.

(5) The State Government may by notification in the Official Gazette and on such terms and conditions as may be specified in such notification, constitute a Committee called the State Fertilizer Committee consisting of a Chairman and not more than 4 other members having experience or knowledge in the field including a representative from State Agricultural University, the Fertilizer industry and Indian Micro Fertilizers Manufacturers Association to advise the State Government regarding the grades/formulations of mixture of Fertilizers.

39. Repeal and saving

(1) The Fertilizer (Control) Order, 1957 is hereby repealed except as respects things done or omitted to be done under the said Order before the commencement of this order.

(2) Not withstanding such repeal, an order made by any authority, which is in force immediately before the commencement of this order and which is consistent with this Order, shall continue in force and all appointments made, price fixed, certificates granted and directions issued under repealed order and in force immediately before such commencement shall likewise continue in force and be deemed to be made, fixed, granted or issued in pursuance of this order till revoked.

Some Specification of some fertilizers are given below:

1. Ammonium Sulphate: Moisture percent by weight, maximum Ammoniacal nitrogen percent by weight, minimum Free acidity (as H2 SO_4) percent by weight, maximum (0.04 for material obtained from by-product ammonia and by-product gypsum) (iv) Arsenic as (As_2Os) percent by weight, maximum	1.0 20.6 0.025 0.01
2. Urea (46% N) (While free flowing): Moisture percent by weight, maximum Total nitrogen, percent by weight, (on dry basis) minimum Biuret percent by weight, maximum 1.5 Particle size -90 percent of the material shall pass through 2.8 mm IS sieve and not less than 80 percent by weight shall be retained on 1mm IS sieve	1.0 46.00 1.5
3. Urea (coated) (45% N) (While free flowing): Moisture percent by weight, maximum Total nitrogen percent by weight, content with coating, Minimum biuret percent by weight maximum Particle size -90 percent of the material shall pass through 2.8 mm IS sieve and not less than 80 percent by weight shall be retained on 1 mm IS sieve.	0.5 45.0 1.5
4. Ammonium Chloride: Moisture percent by weight, maximum Ammoniacle nitrogen percent by weight, minimum Chloride other than ammonium chloride (as NaCI) percent by weight, (on dry basis) maximum Omitted	2.0 25.0 2.0

5. Calcium Ammonium Nitrate (25% N)	1.00
Moisture percent by weight, maximum	25.0
Total amrnoniacal and nitrate nitrogen percent by weight,	12.5
Minimum (iii) Ammoniacal nitrogen percent by weight, minimum Calcium nitrate percent by weight, maximum	0.5
Particle size-80 percent of the material shall pass through 4mm IS sieve and be retained on 1mm IS sieve. Not more than 10 percent shall be below 1mm IS sieve.	
6. Calcium Ammonium Nitrate (26% N)	1.0
(i) Moisture percent by weight, maximum	26.0
(ii) Total Ammoniacal and nitrate nitrogen percent by weight minimum	13.0
(iii) Ammoniacal nitrogen percent by weight, minimum	0.5
(iv) Calcium nitrate percent by weight, maximum	
Particle size-90 percent of the material shall pass through 4mm IS sieve and be retained on 1mm IS sieve. Not more than 5 percent shall be below 1 mm IS sieve.	
7. Anhydrous Ammonia	99.0
Ammonia percent by weight, minimum	1.0
Water percent by weight, maximum Oil content by weight, maximum 20 ppm	

41. IMPORTANT DEFINITIONS OF SOIL SCIENCE

- **Straight fertilizers:** Straight fertilizers are the fertilizers containing only one plant nutrient e.g. Ammonium sulphate, Super phosphate, Muriate of potash.
- **Starter solutions:** The term refers to solutions of fertilizers, generally consisting of N:P:K in the ratio of 1:2:1 and 1:1:2. These solutions are applied to young vegetable plants at the time of transplanting to help plant establish.
- **Soil fertility:** Soil fertility refers to the inherent capacity of a soil to supply nutrient to plants in adequate amount and in suitable proportions.
- **Soil productivity:** Soil productivity refers productivity is the present capacity of a soil to produce crop yield under a defined set of management practices. It is measured in terms of the yield in relation to the input of production factors.
- **Soil conditioners:** Soil conditioner are chemicals which are added to maintain physical condition of the soil. Such chemicals are polyvinyles polyacrylates, Cellular gums, lignin derivatives and silicates.
- **Soil amendments:** Any material, such as lime, gypsum, saw dust, sulphur or soil conditioner, that is added to the soil to influence plant growth by changing the soil reaction or pH or by improving the physical condition of the soil. Fertilizer is also an amendments but the term is used most commonly for added materials other than fertilizer.
- **Band placement:** The placement of fertilizers in the soil in continuous band or narrow ribbons in between the plant rows or around each plant. The fertilizer

applied in band is covered by the soil but is not mixed with it. Phosphatic fertilizers are normally placed in bands to reduce phosphate fixation.

- **Carbon : nitrogen ratio:** The ratio of the weight of plant organic carbon to the weight of total nitrogen in a soil or in an organic material . C:N ratio of wheat straw is nearly 81:1 while that of soil is 10:1. When undecomposed straw with a higher C:N ratio is applied to the soil, its C:N ratio is reduced through bacterial decomposition. To speed up bacterial decomposition, nitrogenous fertilizer is added to the soil at the time of turning of straw.
- **Cation:** An ion carrying positive charge of electricity. The common soil cations are calcium, magnesium, sodium, potassium and hydrogen.
- **Ion:** Electrically charged particles. Negatively charged are called ion. As used in soils, an ion refers to an electrically charged element, resulting from the breaking up of many substances like acids, salts and alkalies.
- **High analysis fertilizers:** The fertilizer containing a high percentage of the plant nutrients. High analysis fertilizers have an advantage in the cost of bagging, handling and transportation per unit of plant nutrient. Such fertilizers are urea (46% N), triple super phosphate (45-47 % P_2O_5) and muriate of potash (60% K_2O).
- **Low analysis fertilizers:** The fertilizer containing a low percentage of plant nutrients, such fertilizers are Chilean nitrate (16% N) and single super phosphate (16-18% P_2O_5).
- **Hygroscopic:** A term used to designate materials that absorb moisture from the air.
- **Equivalent acidity:** A term used to express the calcium carbonate equivalent of acidic residue left by fertilizer materials.
- **Equivalent basicity:** A term used to express the calcium carbonate equivalent of basic residue left by fertilizer materials.
- **Denitrification:** The process by which nitrate or nitrite in the soil or in farm yard manure is reduced to ammonia or free nitrogen by bacterial action. The process results in the escape of nitrogen into the air and is, therefore wasteful.
- **Nitrification:** The formation of nitrate and nitrite from ammonia (or ammonium compound) or in soils by microorganism.
- **Mixed fertilizer:** Mixed fertilizers consist of individual or straight fertilizer materials blended together to permit application in the field in one operation. Mixed fertilizers supply two or three major plant nutrients. The percentage of these nutrients are expressed as fertilizer grade like 10-5-5.
- **Complex fertilizers:** The commercial fertilizers containing atleast two or more of the primary essential nutrients. When such fertilizers contain only two of the primary nutrients, they are designated as incomplete complex fertilizers, while those containing all three primary nutrients are called complex fertilizers.
- **Closed formula mixed fertilizers:** The fertilizer grade is disclosed on each bag of such fertilizer mixture, but the ingredients or straight fertilizer are used in

formulating the mixture are not disclosed. In India, fertilizer mixture sold to the cultivators are usually of the closed formula type.

- **Open formula mixed fertilizers:** Fertilizer mixtures of which the composition and the complete makeup are disclosed by manufacturers. Thus the fertilizer grade, the straight fertilizers and the filler used are disclosed on each bag. Cultivators and extension workers are in a better position to know the quality and quantity of nutrients contained in such a fertilizer mixture.
- **Soil erosion:** Removal of top soil by different agents. Causes: Deforestation Overgrazing. Faulty method of Agriculture. Erosion by rivers. Removal of top soil (for bricks, pots, tiles etc).